Proprieties and Vagaries

Proprieties
and Vagaries

A Philosophical Thesis from Science, Horse Racing, Sexual Customs, Religion, and Politics

Albert L. Hammond

Lanphier

The Johns Hopkins Press: Baltimore

© 1961—The Johns Hopkins Press, Baltimore 18, Md.
Distributed in Great Britain by Oxford University Press, London.

Printed in the United States of America by
The Haddon Craftsmen, Inc., Scranton, Pa.
Library of Congress Catalog Card Number 61-13245

*This book has been brought to publication with the
assistance of a grant from The Ford Foundation.*

To Ormond and Susan, who, without asking why, put aside school-work review and so let this book have needed hours.

Preface

C. S. Peirce's Essay "The Fixation of Belief," which
became a chapter in textbooks on "scientific method"
and a discourse in many classes, sets up three extrascientific methods to demolish: tenacity, authority, and
"what is agreeable to reason." I have been fond of de-
fending them. It is not hard to do. The scientist is also
a man, not only outside but also inside his science-mak-
ing, and tenacity is an aspect of courage, which all
action needs. With his hardheadedness he needs some
humility, docility, and willingness to accept from others.
Along with his prejudices in favor of his own beliefs,
his family, people, and land, he needs some piety toward
tradition, even toward some internally unquestionable
authority like an umpire, a Supreme Court, a *vox populi*,
a faith, a God—even if he calls it just a postulate. And
the "light of nature," *bon sens*, intellectual intuition,
something like self-evidence he must have not only as a
man, I think, but also within the very secret of his knowing. But this is a subtler story.

Furthermore, Peirce rests his case upon an assumption—that the mind works only to get rid of the un-pleasantness of doubt—which I decline and which seems distinctly not Peircian. Dewey may have felt and believed this, but most of the time Peirce did not. And as I reread his essay I feel, in some of it, a note of self-satire, a repressed savagery, which is not un-Peircian and which is felt in moments of frustration by pursuers of the "wonder that is the beginning of philosophy" and science—and which in itself is the answer to Dewey's strange belief that the man with practical problems does and should run to theoretic problems as properly easier refuge.

Nevertheless, when I have said all this, I go on to say that the three rivals—tenacity, authority, self-evidence—are often intruders into the scientific household, are pilferers, burglars, rummagers and disarrangers, arsonists. Not the only, but the most constant, multiple, intricate way they operate is as propriety—the gentle rain and atmosphere of logically composed but not logically directed fashion, acceptance, taste, choice—which influence not only the behavior and furniture of everyday but also our theories, even the most advanced science. In one way or another each of the chapters in this volume has to do with this theme.

It is not a theme of Peirce's essay. So far as I can find there are three one-sentence intimations of it. "Conceptions which are really products of logical reflections, without being readily seen to be so, mingle with our ordinary thoughts." The method of what is "agreeable to reason" "makes of inquiry something similar to the development of taste; but taste, unfortunately, is always more or less a matter of fashion." "But when I come to see that the chief obstacle to the spread of Christianity among a people of as high culture as the Hindoos has been a conviction of the immorality of our way of treating women, I cannot help seeing that, though governments do not interfere, sentiments in their development will be very greatly determined by accidental causes." But there is nothing so frequently repeated, not only in words but also in feeling and action, as the propriety (and this is the method of tenacity). There is nothing that imposes more authority than "what is done." And what are we more "inclined to believe" than the proper?

Why do we make it a point of pride to be hardy and uncomplaining against the discomfort of cold, but vulnerable and indignantly sorry for ourselves when we are uncomfortably hot?

The only one of these chapters previously published is the next to last, "A Defense of Horse Racing," which appeared in *Plain Talk* for March, 1929. I include it because I like it and because *Plain Talk*, a vigorous (perhaps too vigorous) magazine is long defunct. Also it represents in some privileged sense the parceling of proprieties not in time but in occupation and dedication. The editor cut it here and there to get it into space. I want to know what Sir Barton did in that his "greatest" race, which I remember only indistinctly, but I have not tried to find any copy of my manuscript to fill it out. The chapter with the title from Virgil—"Too greatly fortunate, if they knew their blessings"—was accepted for publication. The editor warned me the journal had a backlog. This was in 1939; and then the war began, the editor died; and the paper presumably was forgotten or discarded; at least it never appeared.

"Symbols," "Good Use," and "The Motion of the Earth" were read to Hopkins gatherings; the first two from not fully written-out manuscripts. "The Motion of the Earth" was offered to a journal and returned with the notation that it misinterpreted the Michelson-Morley experiment, which was directed to the existence of the ether, not the motion of the earth. I think this can be shown not to be so; and I think the editor's thesis —in essence it is an orthodox one—itself began as a rapid and undeliberate misinterpretation not only of the significance but also of the intention of the experiment. Indeed it is a neat example of the proprieties involved, and thus adds to the relevance of the paper to a part of its thesis and to the general thesis of this book. I have added a couple of paragraphs to indicate my view of this. At any rate the relevance of most of the paper to the thesis is unaffected even if the editor's interpretation of the Michelson-Morley experiment is partly or quite right—which of course I am aware it may be.

"Bridge" made the rounds at the beginning of the thirties about the time contract bridge was replacing auction. It came back as too long for those that liked the topic and as the wrong topic for those that had space. I think it belongs here, as contrasting the proprieties of three of the world's enduring games.

"Symbols" and "Good Use" have been fully written out for their inclusion here. The first two chapters are written for this place. The second was indeed extracted from the midst of the first: as differing in tone and as being a fragmentary treatment of a topic I hope presently to treat at full length.

The oldest writing here is "Idols of the Twilight." I found it stowed away with this title, forgotten but then recalled as a piece of graduate-student days. It already deals in part with the theme I now come back to. Clearly I was young. I have left out a fair portion, added one paragraph (the second from last), but refrained from rewriting. I would not now write it but I cannot say, rereading it for the first time in some forty years, that I would deny anything there asserted. My more technical views have certainly changed more than these. Perhaps this means something about proprieties.

In offering an assortment of work ranging over forty years, I should like to express my gratitude to some of my teachers, many of whom have also been my colleagues and friends. My first teacher of philosophy, and one of the great teachers I have known, was Dean Edward Herrick Griffin. Then there was my chief teacher of philosophy, Arthur O. Lovejoy, still, I am glad to say, my teacher and my friend. No graduate teacher was kinder to me than James Wilson Bright, in

whose English department I began and in whose seminary I continued long after I moved to philosophy. Two
several-times-visiting philosophers, now gone, were especially good to me, then and afterward: Morris R.
Cohen and William P. Montague. Henry Slonimsky,
known to students as the man who could weep elo-
quently over logic, was the first of my teachers or
colleagues not far from my own age, and my good
friend. When he left came George Boas, whose learning
and wit have entertained and influenced many and
whose kindness to me was and is unfailing. I am in-
debted, too, to later visiting professors, notably Aron
Gurwitsch, Albert G. A. Balz, Herbert W. Schneider,
Sterling P. Lamprecht, Everard W. Beth. I list some
other names, each of which calls for much more than
listing; their companionship has rejoiced me and they
are still about me: Victor Lowe, Ludwig Edelstein,
Kingsley Price, Maurice Mandelbaum. Of the many
from other departments at Hopkins in whose debt I
stand I name just two no longer here: Raymond D.
Havens, and Leo Spitzer. I like to add my two long-time
Goucher friends and colleagues in philosophy, Gertrude
Bussey and Raymond P. Hawes.

Not so much my teachers as my constant benefactors
have been the directors of the night courses and the
summer school: Edward F. Buchner, Florence E. Bam-
berger, Robert Bruce Roulston, Francis Horn, Richard
A. Mumma.

I can look back to undergraduate Hopkins and name

some teachers I remember especially happily: John C. French, Herbert Eveleth Greene, John Martin Vincent, Joseph S. Ames, Wilfrid P. Mustard, George E. Barnet, Kirby Flower Smith, John Hollyday Latane, Lorraine S. Hulburt, Jacob Hollander, Ronald Abercrombie.

I can look back of college to Miss Ida Jarrett, of about the fourth grade, and Edward Raymond Turner, my first man teacher with my first all-boy class in the eighth grade, a potent experience which he made so from the first day when he addressed us as "gentlemen" (he was later professor of European history at Hopkins); and at City College Herr Raddatz, Ernest J. Becker, Alexander Hamilton, John Morgan, Alonzo Smith, Frank Blake (who also later, in the Oriental seminary, taught at Hopkins).

My chief fellow student in philosophy at Hopkins is still my friend in Baltimore, Alvin Thalheimer.

My students, graduate and undergraduate, are, I think, my chief original benefactors in philosophy and my rewarders later. One stopped me recently and quoted something from a lecture of, he said, 1921. By such is the spirit made glad.

I add a word of thanks to my single academic home, the Johns Hopkins University, an institution singular in more important and better ways than that.

A. L. H.

BALTIMORE, MARCH, 1961

CONTENTS

Proprieties and Vagaries

I

Proprieties
and Vagaries

ONE TROUBLE WITH the relativists is that they are almost never relativistic enough. They come to a preference, a prejudice, a whim, a habit, a taste, something welcome to them as subjective; and they stop. If they would push on they would get out. All these subjectivities, almost always, are themselves relative to a beyond, they have their causes and usually their reasons. Our values (in that one of the many uneasy senses of "value" which points at the rules of our preferring—known or unknown, expressed or unexpressed, rigid or sporadic) are partly from chance, partly from cause, and partly from reason. And so far as they are from reason they are asserted on a basis of perceived fact—facts inside our skins or outside, descriptive or valuational. And the "reasoning" is deliberate or implicit, logically good or bad, and our own, someone else's, or our group's or our time's.

I am an anti-relativist or absolutist here interested in certain relativities, where what we accept—in high science or in trivial habit—is determined by group proprieties. If I am concerned most really with the epistemology and axiology of the situation, I am concerned more immediately with certain instances, now and in history.

One advantage of seeing an absolute behind a relativity, as against seeing relativism as absolute, is that the first gives a way toward understanding not only the act and the actor but their correctness; the second, proud of understanding the act and the actor in terms of each other, can find nothing beyond in terms of which to understand them. The first gives hopes of criticism, even reform. The second can, or should, only accept. I am a preacher, a defender of the currently indefensible, a critic of the respectable. A full relativism should find even absolutism perhaps false but unobjectionable.

In the big classroom down the hall this summer several undergraduates—three today, I think—are working on the supplies of textbooks that keep piling in for the fall term. They have taken to singing, chiefly "The Volga Boatmen," and their untrained voices (they are not glee clubbers) sound very good. I have seldom heard a voice, especially a young voice, used at all naturally that did not sound good. Some are better, some cannot keep on key, but, as such, human singing is pleasant for a human to hear. One is not apt to sing when he is in a bad temper. Yet after a while, the singers down the hall evidently remember they are Americans and take it for granted that all non-professional singing is a joke at best or a suffering to others. Then they caricature their singing, and laugh at it apologetically and derisively.

I suppose this strong tradition is in part a hang-over from the period of strict professional music (apart from some German or Italian immigrant influences) up to about 1910. There were Jenny Lind and Caruso and there were those who listened. But this all changed. Then there were the dance bands and the jazz sessions. But singing, the more natural music, did not shake as loose from the tradition as instrumental. It did spread greatly in the sorts of singing listened to, in the time given to listening. Today many spend much of the day and night listening to singing on radio and television, and obviously there are many singing on radio and television and with bands, without on either part any feeling of its being a joke or an offense. But the singing is still professional, by a "singer," set apart. Let someone not before a microphone "threaten" to sing and there is the ritualized outcry of alarm. Yet many of those applauded when heard over the airwaves are about as untrained and unselected as those derided. A Ricky Nelson can be on TV just because he was born into a family with a non-singing family TV program. At the appropriate time, "when the beard appears," he runs into a script that calls for a song: Ricky twangs a guitar and sings to his girl and makes a fortune. And I am for it, I like to hear him. But

3

I also like to hear the boys down the hall sing as they work at the piles of books. I am not urging them to go on radio. Ricky is probably better. But they, and most of us humans, even Americans, are good singers.

If singing held more tightly to this tradition than instrumental music, we can say it was more tightly bound to begin with, and that singing stayed more professional from separate causes; and we can also say that there is something especially lasting about a joke and that personal depreciation (when it is not felt to be greatly important) has its natural motives. Playing the piano is obviously by learning, not nature, and we cannot well find a joke in playing badly or make fun of ourselves for not being able to play well. We may not be born brave, but we are not likely to make easy fun of ourselves as cowards. We all have voices; the professionals have good voices; the rest of us can make fun of ourselves and all our friends for having intolerable voices. But this is not to say any such tradition had to arise or stay. Most places and most times have not had it. We do. I have kept tab for years and I have yet to hear a gathering of more than two Americans when an intimation of an imminence of singing by one of them is not followed by the dutiful outburst of the deprecatory humor of the tradition.

For awhile this summer there was a cartoon on the bulletin board in the Baltimore *Sun* newsroom. I do not know who put it there and I did not notice who drew it, but it was well drawn and printed. My description is not one of perfect recall but it is not overemphatic—the clear intent of the cartoon was to be emphatic. Its caption was: "The day the copyboy was caught whistling in the newsroom." The culprit is departing, dragging his few belongings and with his head as nearly between his legs as a walking human can get it. The managing editor is standing in the door of his cubicle, fists shaking above his contorted face. The city editor is pointing sternly at the door with one hand and waving away with the other. Two copyreaders have crawled under the big horseshoe desk covering their ears with their elbows. A reporter has a chair over his head and a columnist is running from the far corner brandishing a pistol.

The muscular violence is excessive but the attitude is represented accurately enough. Especially a home of traditions, the newsroom (I am not sure of the rest of the building) has few if any traditions firmer than that whistling is out. I have tried singing and been met by careful inattention or by the proper protest of humor (although, in the full propriety, that is supposed to be begun or invited by me) and been ultimately silenced by circuitously delivered serious intimations of scandal. Even I would not dare whistling.

Yet, also in part by tradition, the newsroom is one of the noisiest and least inhibited of places—and least tolerant of intolerance of confusion. The wire services and dozens of typewriters chatter and the teletypers go on long periods of ding-a-ling which no one pays the least mind to, telephones erupt in all directions which almost everyone leaps to answer, two or three shout or scream "copy" and boys run or crawl in response or play ball across the room with a wadded ball of paper and do not respond, the managing editor calls to the city editor and the news editor, at intervals or in series a pneumatic tube hurls receptacles at a crashing mark—and all this is broken disquietingly by bits of silence. But whistling there is not.

I have been assured this is reasonable. Whistling has a rhythm; so too does the thinking of these noisy intellectuals. Head-writing especially is rhythmic and the external rhythm of whistling breaks up the internal. This is a bit preposterous. People differ psychologically and physiologically in their annoyances and indifferences. (I abhor drumming when I am trying to work and drumming is more obviously rhythmic but most copyreaders do not mind it.) The whistling sound is not as normally pleasant as the singing sound and can become a shrill pain in a small hard-walled room. I know newspaper men who have no objection to whistling anywhere else and some who start to whistle as soon as they leave the newsroom. It is preposterous that whistling should be insupportable in that one place, but I think it is quite true for newsroom denizens. They really cannot stand whistling. But most of them could and would if they had not learned this propriety.

Coming up town in the bus yesterday I was thinking

of how to carry on this argument when, the block after I had got on, a girl got on with a pretty little transistor radio, clear with songs. I said to myself, "How can anyone so blithely commandeer the time of people who had hoped to make a chosen use of the twenty minutes' ride up town? I suppose it never occurs to her that one can do something while getting from one place to another. This is why people fail to see what a time-waster the automobile is." But I stopped this familiar trickle of rumination and docilely gave my attention to the song. "No matter what I do, I can't forget I love you."

And then a more particular thought suggestion came to me: "This is how to carry on that argument; by reverse. Here you are prevented from thinking by a song. Is not this what you just called a bit preposterous?" And I went back to listening to the song. "The memory of your kiss."

But in scraps through the dutiful listening I was thinking: "This is one of your own proprieties. The way the transistor interrupted and hampered your task, to be sure, seemed to, and in part did, but did not altogether, controvert your thesis about tune in the newsroom. But it further illustrates your general thesis of how much we are governed by intellectual proprieties. If it laughs at you, it applauds your contention. It is not physical compulsion of rhythm, not psychological love of sorry-for-myself ballads, not esthetic dominance of artistry that interrupts your intention—not even the lazy willingness to take any excuse—not something natural but something fabricated. It is a long training and acceptance of the impropriety of not listening to music when played for one—even by a machine. These are the manners of those days when music was more officially a presentation. This is the special training given by your sister, the pianist, who would crash both open hands on the keys, get up and leave the room if someone made conversation while she played. This is your own habit of tolerance but nonimitation of the younger generation's habit of conversing while the radio or television is on and the habit of the still-younger generation who study with record player added to broadcast. You have been known, twice, as a visitor, when someone asked you a question by name, quite undeliberately to reach and turn off the radio before answering—and apologizing.

"But this is always of music in the atmosphere of continuing presentation. I do not think it would apply to the bits sung to himself by the passer-by, the workman, the copyboy in the newsroom; unless natural reasons of beauty—or loudness—intervened. Still let this be a contradiction of attitudes. It would not be singular. One ingrained propriety, derived and in part intellectually derived, drives the newsroom frantic when a boy whistles. My withers are unwrung. But another propriety keeps me from talking, reading, or working when a pianist or a sound track plays; although my newsroom friends talk or read the while."

Friday, Saturday, Sunday, Monday were to me the most uncomfortable days of a Baltimore summer which

8

had more than the usual number of uncomfortable days. Monday afternoon the weatherman gave up and rubbed it in by telling the city there would be no relief for five days. The Bermuda high (Baltimoreans know that term well, and badly) was reinforced and there were no cold fronts to the west to threaten it—just a weak low up north sliding out to sea above the clockwise drift around the high.

When I came home from work that night at twelve-thirty it was cloudy with a bit of cool east breeze. Thunderstorm, I thought. But at eight the next morning it was solidly clouded with the east wind blowing. At ten the sun appears and I think "Here we go back in the southwest trough." But I go out and look up. "And now the storm clouds fall apart,/And now the west winds play." Northwest winds, with little fleecy clouds moving fast and steadily from a little north of west and the air is dryer and colder. I know the high is moving and Canadian air is on the way—slowly perhaps and perhaps the high will fight back but—we are already in a different climate, outside and in. "And now the storm clouds fall apart,/and now the west winds play;/And all the windows of my heart/I open to the day."

The August sun is still potent and most of the air is still Gulf air from the southwest; and as the afternoon temperature goes to ninety degrees people look as desperate as the day before. When I cheerfully inform my friends at the University and at the *Sun* office that the climate has changed and help is on the way they say I am playing with them and making it worse. In the noon paper the weatherman agrees with them, just as he does in the two o'clock paper. And I say the trouble with the weathermen is that they never go out and look at the sky and feel the air. At three o'clock the Weather Bureau announces to the public that relief is promised. My friends allow themselves to be hopeful and I assure them they live in a world not of perception and feeling but of sheer conversation and that this is what Aristotle meant when he defined man as a rational animal. He said that means man is capable of learning grammar. He should have added it means that man can and normally does learn a whole stratification of acceptances that get between him and what he might directly perceive and feel as well as what he might Socratically and scientifically discover; it means that man can make a respectable social automaton of himself, can with the debris of thought escape the chore of thinking, can make a fool of himself, can make himself an eccentric, a fanatic, a scientist, a saint.

I am not saying this represents any material tradition as to Baltimore weather—there are such, certainly —but that all this represents a bit of the conversational way in which we not only talk but feel and see; a way, a defect of our essential virtue which makes the chief basis for the promulgation and maintenance of traditions and proprieties.

There is a first-person addendum for the critic, the clear observer, the esthete, the Socratic, the scientist. I have sometimes done as above with the weather (and other things), the weatherman has stood out, and in the upshot the weatherman has "proved right." Have I made a fool of myself with some acceptances of my own? Have I merely made a mistake in my observations, feelings, decisions, as a result of bad physiology, bad logic, or chance? Has the weather fouled up probabilities? All these happen. But the ineradicableness of the chance that it is the first—my own proprieties— should keep the critic of proprieties sober.

A rule is a sort of propriety; and the more proper propriety—more proper as less consciously adopted and used, with more of the tradition and less of the instrument—often grows out of rules. A rule I take to be a maxim to guide action, especially action that must be fast, in the light of some more important or more frequent end. But since it guides action it works in a concrete and mixed-up world. The less frequent end may be the relevant one or the less important may be, in the circumstances, the more important. How often or easily, if at all, the rule will give way will depend upon the subject matter, the sort of rule, the rule user. Much of the hard-to-define difference between parties in the long dispute over relativism in values lies in our habits as to rules: swear by them or at them. That rules have deserved imprecation is clear. That we cannot do without them is clear.

The printer, the composing-room worker, is a very fast

worker. A city newspaper can pull out the front page, put in a new story with a new line and with new kinds of heads on most of the stories that remain, and be out in less than ten minutes, the page clean. The compositor, just one link in the chain but the mechanical key to it, has no time for debate or weighing. He has a quotation mark and a period. The period goes inside the quote. The writer and the copyreader may know that in logic it might be better the other way. They will hardly get it so. You say the compositor should follow copy but to follow copy with those little points would itself slow the eye, the hand, the edition. Anyway the smaller point looks better framed by the larger quote; and the printer is one of the true estheticians of the modern world. So he has a natural basis both in economy of time and in good looks for his rule; few rules in technique, art, business, or morals can do as well. Yet it is clearly at times in the wrong.

From such rules other things grow. Among them is the tradition, fairly firm among makeup men and compositors, that the composing room knows more than the newsroom can ever hope to learn—as indeed, on this ground, it does. Is it not a wider attitude, not only a tradition but an understandable inference from false premises, that the one sure mark of knowledge in all fields is promptitude and finality in answering any question?

The rule to write from the top of the page down is doubtless based on good physiology and the physics of writing instruments which draw better than push and even more on the desirability of keeping what has just been written from under the writing hand. One result is that, when Aristotle or the Academy got around to teaching us to classify, the genus is "higher" than the species. Then we note the fact that the putting down of names of individuals according to rank or the putting down of the names of the ranks of a hierarchy—say the celestial one of the Pseudo Areopagite and Milton or the United States Army—has a resembling pyramidal structure, and we get the variously influential tradition that the more general is the better and that a classification is a hierarchy. This tradition lingers on despite the curious but well-known fact that only existent persons or organizations can "command" and all these are outside the whole classification except in the sense that they are members of the classes—and in this sense they are all the members of classes all the way from the lowest to the highest. It would be confusing in the Army if the only members of the class of generals were those who at the same time were privates, lieutenants, and colonels.

A buttercup improves in the glory of its golden color as it withers. It also becomes more fragile, but I have several times kept a buttercup in an unused lapel from one year to another. But it is a rule that withered flowers be discarded, and I have had kindly females with a regard for my propriety pinch my improving buttercup from my lapel with an admonition for my absent-mindedness. Well, most flowers do wither poorly. I suspect more of the force of the feeling is that as automobiles, let us say, wither they no longer serve their prestige purpose.

On August 1, 1959, the Associated Press sent a dispatch from Urbana, Illinois, which began:

"Are you a person with beliefs that stand like Gibraltar against all opposing arguments? Not so, unless you have developed a defense against persuasion, says a University of Illinois professor who has tested brainwashing techniques on 200 volunteer students."

The beginning of the second sentence confirms the suggestion of the first that it is being taken for granted that everyone wants to be a person whose beliefs are impregnable to all argument. The second part of the second sentence, however, makes some amends as showing what the author is thinking of: not just beliefs but proper or patriotic or our-side beliefs, and not argument but brainwashing.

The Baltimore *Sun* carried the two sentences: "Are you a person with beliefs that stand against opposing propaganda? Not so unless you have developed a defense against 'persuasion,' says. . . ." The headline was "Uncriticized beliefs prove easy prey to propaganda" in one edition and "Criticized beliefs prove firmer against propaganda" in another. The Socratic echo was proper, if a bit ironic, for the dispatch went on to tell how Professor William McGuire, 34, with his two hundred volunteers discovered (without any hint, on the

part of the AP at least, that there was any rational antecedent probability in the outcome or that anyone, notably Socrates, had ever said so before) that, of three
groups agreeing that yearly medical checkups are good,
the group with the thesis "unexamined" showed the
most defections, those with it half examined (given
favorable arguments first) came next, and the group
which examined the thesis pro and con showed the least
defections when propaganda methods against medical
checkups were applied. Even right opinions, says Socrates,[1] "while they abide with us are beautiful and fruitful, but they run away out of the human soul and do
not remain long, and therefore they are not of much
value until they are fastened by the tie of the cause."
The dispatch gave no hint of any distinction between
argument (and perhaps proper persuasion) on the one
hand and improper persuasion, hidden persuasion,
propaganda, pressure, brainwashing on the other.

But the distinction and the faith in the rational side
are fundamental to Socrates and to our present Western
side against the East—or to what was our side. Have our
opponents brainwashed us into becoming brainwashers?

In our talk about Plato's throwing out the artists, we
post-Romanticists are usually content to disapprove, to
note the paradox of Plato the artist against the artists,
and to say that the tenth book of the *Republic* is not
Plato's only word on the subject. Indeed, it is not, for he
defends art and denounces the danger of art, and we fail

[1] *Meno,* 98a.

to see the seriousness of the charge, at least the seriousness with which Plato took it. Discourse and all imitation are under the command of truth. Art is constantly pulled to forget the command in order to be pretty or entertaining or profitable to some other end, to be a teacher of bad manners when it is in the first person (*Republic* III), to be "cookery" (*Gorgias*) or to become propaganda and brainwashing (*Gorgias, Phaedrus, Theaetetus, Republic*).

"Persuasion" is ambiguous. "Argument" is honest and on the right side (*pace* the AP) but is apt to be taken too narrowly, intellectually, argumentatively; it is too lawyer-like, and Plato disliked the lawyer, too. The Sophist had been smart (not wise) enough to turn argument itself to propaganda and "make the worse appear the better reason." So, a bit curiously perhaps, Socrates, I think, began with, and Plato came to, an awareness that something more than argument must be called to the service of truth. The old word "reason" (not "reasoning") may do. If the irrational beyond reason in the direction of dishonesty and of ulterior service, distortion for a purpose, is the foe, then something in the direction of honest perception of value— feeling for fitness, faith, inspiration, common sense, sagacity—must be a part of or the ally of reason (*Ion, Phaedrus, Symposium, Meno*).

Why not suppose the world purposely purposeless?

Any good gambling house is. And its purpose (too?) is to separate the winners and losers.

This is not to make it all simply luck. Gambling houses play poker and take cuts on the horses as well as offer roulette and chuck-a-luck. Nor is success at the latter nothing but luck.

Luck, indeed, at this point (I want it later) is beside the point. If the merits of the participants are to be sifted by conditions and events, by the world, the world must be purposeless, on purpose.

But if it is only merits as given facts which are to be sifted, presumably the Creator could do it more directly, easily, quickly. But suppose merits are not only to be sifted but also to be developed, encouraged, discouraged, created—that is, tried. Then the world, through natural law and conditions of birth and deliberate judgment and courage and effort, and luck (as chance, not fate of any sort), and freedom, real freedom —serves its purpose if and only if in its own unrolling, its inner determination if any, it is purposeless.

Evidently a purposely purposeless world is not a completely purposeless universe or a purposeless ontology. It allows, and the motive of my story calls for, purposes in the individual. I would say all the souls—humans, animals, angels, ghosts, and what not, if any—have purposes: fragmentary, deliberate, habitual, more or less enduring, more or less consistent and overarching, more or less intense, wise or foolish; for to be or have a soul is to be in some degree aware of one's situation in fact and value, to have choice, to be possibly free; and this is purpose. Also the purposely purposeless world allows, and the phrase calls for, a purpose before or behind or after the world: a purpose on the part of God in creating the world, if you prefer; or, if you prefer, a "final" purpose discernible in or attributable to this natural system—attributable in emanation, evolution, or sheer chance. Here the world is and thus it works and is worth while. The world and its events in their own unrolling are the falling out of indifferent natural law and indifferent chance, in the midst of which the animals' purposes succeed and fail; and all this in its general purposelessness works to a purpose which itself may very largely, although not altogether, fail.

And it is the intermediate purpose here denied which is usually in question when ordinary people wonder, ask, lament, or sneer as to the purpose of the world. To be sure the denial of God's purpose is often associated with these responses; but people are likely to derive this denial from their failure to find any purposiveness in the world or they are content to dismiss the possibility of an outside purpose as that of a purpose not good, or opaque, or irrelevant to us. And it is notable that the general arguers for a purposeless world are not apt to insist on the nonactuality of particular purposes in animal life. They are willing to think of a chess player, a lover, a revengeful murderer as having purposes. And they are simply not interested to deny that their own purpose is to find out the truth, to be bravely honest and free from wishful thinking, to put up a good argument, to be in fashion, or to show off.

"Purposeless universe," meaning a sum total of being

in which there is nowhere—inside the material world or out of it, in God or in any animal—such a thing as purpose: this is surely not an unassertable or unasserted proposition. But students' papers and discussions have taught me that arguing against such a total purposelessness does not meet their point or satisfy them even in opposition. What they mean is that the world of physics does not show any purpose in its working, that the world of biology does not show any purpose of a creator, that the world of history does not show any achievement of general purpose, that all the rivers run into the sea, yet the sea is not full and that this too is vanity and vexation of spirit. It is not usually realized that these are different theses and that adding them together does not cover all the ground apparently meant in the general text "This is a purposeless universe" or even world.

The strength of the argument that runs into the feeling comes largely from physics. It should be noted, however, that it is suspiciously safe here. The modern physicist explicitly lays down the rule that anything not in his purposeless categories is, at least, not physics.

The nineteenth-century biologist usually tried, and present ones too try, perhaps in a somewhat different way, to take the same road as the physicist; but not all. And even those who do are apt to agree that the language of adaptation to ends (of the epiglottis that closes the windpipe as the food goes down, for instance) is too insistently useful (itself adapted to its end) to be barred or eschewed, and that this may be allowed, without full scientific indecency, to those who want the realism of a sort of Aristotelian immanent teleology. But biology is below the level of conscious intentions and so we do not get much of the purpose of the chess player (but some of that of the lover). And the sort of general or biased utilitarian purpose our dismal or superior denier is denying is not much spoken for— seems often spoken against—by the well-known "nature red in claw and fang."

The historian sometimes attempts the anti-final-cause definition of the physicist. But then too, in the zeal to be "scientific" which usually goes with this motive, these historians are apt to have their laws, and laws of history not just laws of physics, biology and so on working in the events of history; and this quest sometimes leads to laws of history which have a sort of purposive design—only these too are pretty sure to be interpreted as frustrating the human hopes of a real accomplishment and negativing any general purpose beyond the normally frustrated particular ones. There may be some workings out of individual psychological purposes but workings out both baseless and truncated because of the purposeless physics and the purposelessly patterned history in a negative theology. Thus, it appears that the feeling which is usually expressed in the denial of purpose is based, on the one hand, on the assertion of natural and historical laws indifferent to us, and, on the other, on a belief in the actuality of our individual purposes, particular, perhaps illusory, absolute. This brings us back to the complaint made at the beginning of this chapter against the relativists as being not relativistic enough, as not seeking the objective grounds to which their subjectivities are relative. It is the thesis here that our purposes as conscious are necessarily cognitive. Their matters and their points are largely from proprieties—from fashion, accident, and intellectual association. But as cognitive they are necessarily rational and logical; no matter how irrational and illogical, they cannot be nonrational or nonlogical. And on the other side the indifference of the laws we assign, or discover, or that are in the world may be relevant and needful to the rationality of our purposes.

It is, of course, in the psychological and literary worlds that this is most apparent: the frustration of actual individual purposes sparks the denial by the individual that there is any purpose in a helter-skelter of seeming purposes constantly defeated or disillusioned.

So the question "What is the purpose of it all?" works through "What is the meaning of it all?" to "What is the use or the good of it all?" To each of these it is well to answer, "Suppose an ideal world in which the question has a satisfying answer; then be clear as to that world and the answer it affords." A world of efforts in which no effort was ever defeated would suggest the query: Why not the same achievement without any effort? And this suggests: What could be meant by an achievement which is the achievement of no effort? A world of unremitting pleasure has been at times indicated in argument; but to a realistic imagination this is apt to seem less attractive (because less active) than Socrates's supposition of pure itching and scratching. If we are told to keep effort on the ground that effort is itself good, as indeed it is, this thesis requires that effort be real, that is that it faces defeat and defeat is possible.

Do we want to go to a gambling house or a race track where we are metaphysically guaranteed all bets made will win? The real gambler in the sense of the possessed one would quickly find his dope had lost its power and would be in the torment of the deprived addict; the real gambler in any more qualified sense would soon be bored or aware of wasting his time. To be sure, the non-gambler, either as the natural insurer or as the man who finds his gamble in less artificial ways than a game, would not want to go to the track or the gambling house anyway; but there is no one, I think, so in love with security as to relish a world without any accomplishment or futurity of openness. I hasten to add we all at times want out. The addict yearns from his enslavement; the nonaddict temporarily possessed yearns with the more hope; the merely worn-out and the beaten-down become almost afraid to think of some bit of release if not of success.

I have twice gone more than seventy races bet without a winner. And this was when I thought—and others thought—I was unusually good at the pursuit of winners; working hard and full time at it—watching every horse in every race for signs of future races, watching workouts and studying the reports of clockers of workouts, poring over past-performance records, attentive for all the signs in the paddock and on the way to the post. Once after sixty-some losers I came to a race— the last race one cloudy day at Thorncliffe, outside Toronto—in which one horse seemed easily best and only one other seemed at all to belong. I put a two dollar bet on each, knowing I would very probably lose and at best could win about a dollar, but I was trying hard to break my streak and cash a ticket. At the sixteenth pole the favorite was running easy by two lengths over the second choice, the rest struggling around the stretch turn. The favorite stepped in a hole and fell, the second choice fell over him; and I went back to the hotel with one more failure. Then one is beyond talk of a purposeless world and confronts the persuasion there is a particular anti-purpose to one's own.

What in more particularity is being sought by the seekers and non-finders of "purpose" or "meaning" or "use"? Sometimes it is, and sometimes they will say it is, what once they believed or thought they believed; and these beliefs are largely the proprieties of times, places, classes, and families, usually beliefs in some sort of Providence. There was the Stoic faith in a general and inherent Providence carrying the whole process of nature; there was the Hebrew trust in the divine favor which, if properly served, would rescue and preserve Israel; there was the Augustinian assurance of the city of God for the elect; there was the belief in miracles irrupting into the world of natural law to enforce the right side at crucial intervals; there was the mystic's world from which the flight of the alone to the alone was at once hard and easy; there was the belief in a special Providence that watched over the fall of the sparrow and beyond our knowing overruled chance, law, and evil intention and folly to make all work for good. There was in general some assurance of some guardian against the "horrid notion of chance" or "implacable natural law" or both in favor of a predetermined good. It is a precious part of the notion of Providence that the specification of what the good end is can be left to the wisdom of the power that shapes the events to it. But surely most of us when we accept any Providence are apt to read our own good—and not without warrant—into that good end and—with less warrant but not without any—to interpret that good of ours in terms of our present purposes. There was and is always the unwarranted hope that in this gambling house the wheels and dice shall be a little but helpfully crooked to favor me or at least the right people.

Well, the right people are safe; but not in the world of events. Three things are among those I think one can be quite sure of. The good man is better off than the bad. If you are "good" in order to get this reward of being better off you are not good and will not get it.

The rain, desired and undesired, falls on the just and the unjust alike.

Sometimes a loss of belief in Providence and the resultant denial of purpose, meaning, and good in the world, is not much more than a loss of "belief" in some race track or gambling house superstition, magical trick to nudge events in our direction. Our denial of the world's purpose is a complaint that we are more on our own than we want to be.

I have no objection to miracles as possible. I think there are some. But God does not commit many. Our different attitude toward miracles is a very real change, partly of fashion, the lapse of a propriety; and more than this. Doubtless we do not altogether see the way with miracles that the eighteenth and much of the nineteenth centuries show us in their printed arguments. But surely, too, they had not come in view of our skepticism as to the need or the cogency of reported and committed miracles as an argument for theism; or of our doubt as to how we might easily or certainly know a miracle to be such if there were one before us. The dialectical proofs of the existence of God retain some flavor of proof for those who like that sort of thing—in general depending upon whether or not one can find any flavor of validity in the ontological argument, to which in the end the others appeal. Our fashions today are more apt to feel force in the more empirical argument involved in the fact that some of us feel that God is the best way to account for the child's love for the father and still more the father's love for the child, to account for the greatness of great music and poetry, for all these earthly dress's "bright shoots of everlastingness."

We know, of course, that the psychologist turns all these around and easily accounts for God in terms of these psychological absolutes, about which we think he is not relativistic enough. And we are sure we are more understanding of and more tolerant of his fashion and theory than he is of ours. And in this area, I think, miracles come back—for the person to whom they happen or seem to happen. When one needs, and seeks, and then has it happen—not with any violation of natural law but in utter violation of any expectability of coincidence—the antecedent need and the answering shift of normal pattern make one willing to say "miracle" or to believe in an occasional intervenient Providence even in the flow of physical events. If conventionally sophisticated, one reminds himself that even the most improbable event can be actual, and if the improbability is purely physical or "worldly" he may let it go at that; but if it is strongly personal and moral he may find it more than so. If this is to have any force for others, if it is to be generalized, it may be in the force which some come to feel in that sort of constant miracle whereby at least some of those who "wait upon the Lord," when "the young men faint and are weary," are able not only to "run and not be weary" but even to "walk and not faint."

But although this be generalizable and may have appeal to some of the unembroiled, it is still the man concerned who is chiefly persuaded. "Still the heart must bear the longest part." The anticipatory foreboding, desperation, questing hope can be in the acquaintance only of the man who says, "Now indeed unless I get help I am done in." Then when he does go on through he is apt to say in something like Aristotle's "practical syllogism," "There was help." Surely some see their helplessness much too easily; and they have the less argument thereafter. "Natural fortitude," the psychologist says; and I agree, but in natural fortitude I see the naturalness of God and His help.

I came in from Jefferson Park, New Orleans, once in a taxicab with four others among whom was a small, neat, gray-haired man with a goatee and a gold-headed cane. I knew he had owned fine horses and had also been one of the country's biggest bookmakers; and that he had been wiped out when Black Gold won the Derby—I suppose the widest slaughter of "books" in American history. He paid off, sold himself out and went in debt when some paid what they could and forgot the rest and when some few simply welshed and disappeared. Someone in the cab remarked on the gameness of a horse that day and the man with the cane said, "People talk about the gameness of the thoroughbred, but I have known very few." Then looking meditatively at the passing countryside he added, "But I have known more game horses than game humans." I have often thought since that he was wrong on both counts—rhetorical but unfair. Animal gameness—not the imaginative courage to leap to the potential but sheer ability to "take it" in imposed need and gameness in unexpected persons and places—is one of the splendid things of the world: too splendid, some feel, to be taken just as given fact, yet not needing explanation so much as explanatory of beliefs and theories. In this and in such is involved the notion of God; so the materialist and the religious man agree: he sees it as the cause and component of a myth; we see it as a rational ground justifying a true belief.

I do not recall whether at the time of the taxicab ride I had seen or was still to see that same Black Gold—most storied of race horses before birth, in life, and in death—then an old horse in a minor race at the New Orleans Fair Grounds, coming past the eighth pole in second place and trying for the lead. His off front leg snapped above the ankle and, standing by the rail, I saw the bone protruding and his foot and hoof flapping. He finished the race, seventy yards or more, and held second, on that bone. They helped him away from the track and the grandstand, held his head in a man's lap, and destroyed him with a strychnine needle. His grave is near Pan Zareta's in the Fair Grounds infield; and every year after the New Orleans Handicap the winning rider with the winner's blanket of flowers, the winning owner and trainer, the judges and stewards go out to the infield, the bugler plays Taps, the flowers are put on Black Gold's grave, and the crowd, which will keep quiet for nothing else under the sun, stands quiet for two minutes. I have seen it several times, and regularly some, mostly men, shed tears. By tradition and propriety a tough place, the track is also senti-mental.[2] For the most part it is simply tradition and simply sentimentality, but there is something more there; and in the respect that the tough and unre-spectable give to the basic goods, courage and honesty, the man who has made a propriety of seeing the divine in the material again finds what he looks for.

We do not think of miracles in the way that, or as much as, the eighteenth century thought of them. In reviewing a book about the eighteenth century,[3] I once wrote a report of the minutes of several meetings of the Almighty and His cabinet of angels. "The folly of the present world was noted and it was urged that some interposition was called for. A miracle was authorized and a committee appointed in the matter. At the next meeting progress was reported in the study of the pro-gram. But, at the third succeeding meeting, the com-mittee reported bafflement, indeed failure. A number of miracles had been put on. It could not be said they were without effect: the scientists had been given great work; great ingenuity had been evoked; and consider-

[2] This is the old track: "on the turf and under it." Funeral proprieties, too, change.

[3] Review of W. McI. Merrill; *From Statesman to Philosopher: A Study in Bolingbroke's Deism;* in *The Review of Religion* (November, 1951).

able changes had been embodied in science. But no scientist had even thought of a miracle, and, if any public man had had such thoughts, he had been careful to say nothing. Only a few who had been very willing to see miracles right along gave any credence, and even these were disposed to prefer their own older list of miracles."

Miracles as "violations of natural law" I would not ask for nor would I rule them out. I have never had any reverence for, never shared the propriety of, the inviolability of natural law. There is always chance. And there are individual purposes which shift and direct the motions of the bodies by use of natural laws. The moral, practical world has its own laws of creation and retribution far more inviolable than those of physics, and here is where we need those miracles, the undoing of the past and grace. (John Foster Dulles was once quoted in the papers as saying: "Church people I respect in moral matters but when they get into practical affairs. . . ." What a curious—or is it?—contrast of moral and practical.) The world of physical events I am willing to leave to matter moving by impact and chance; and I think there may be, and is, a purpose in this as there is in an honest gambling house, where all the dice and roulette balls and cards move by indifferent mechanics and chance. Some players go home winners, some losers; some of each party go home better than they came, some worse. The house will give you a ride back to town and, if a good house, will give you a five dollar bill if you went broke—to buy you dinner and get you home. I like to think the universe is as kindly—perhaps a miracle, indeed: another chance really. But kindliness by itself is not enough, is not creative in existence. Can free virtue be created incidentally and from without? Or can it be created all at once and as necessary? The good spirit, it would seem, if it is to be created must also create itself, in freedom and in a world of indifferent events.

Perhaps our own particular gambling house is not altogether honest—is honest in its mechanics but not in all its attendants. "In the beginning God created the heavens and the earth"—or was it Satan who created the earth, Satan and the others after the expulsion from heaven and as a hopeful way of getting back at God? Perhaps Milton, our best authority on the details, was just a bit misinformed by his intelligence system. Satan and all his hosts, "hurled headlong flaming from the ethereal sky/with hideous ruin and combustion," did on recovering meet in conclave on Pandemonium. And after Moloch, "horrid king," urged outright war, which all had had enough of, the "lewd" Belial, "industrious to vice," and Mammon, "the least erected spirit that fell," made waiting and apparent peace seem not only wanted but of some promise, and so prepared the temper of the assembly to be seized by Beelzebub, next to Satan "in power and next in crime." We are told he said:

31

There is a place
(If ancient and prophetic fame in Heaven
Err not)—another world, the happy seat
Of some new race, called Man, about this time
To be created like to us, though less
In power and excellence.

Was Beelzebub's scheme, prompted by "the author of all ill," led up to by Belial and Mammon, was it to raid this God-created world or was it more radical—that hell itself should create a world and men and then "seduce" it and them to the "party" of the "bad angels" and so "interrupt His joy" and build their own strength? "The bold design/Pleased highly those Infernal States and joy/Sparkled in all their eyes: with full assent/They vote." Belial and Mammon, lust and greed, promised to be constant and ingenious in seduction of the new creatures. All the others foresaw ways of helping to hurt. Satan himself took off on the "dreadful voyage" through "unessential night" to begin the work of creation. Surely the promisers of temptation and misleading have been hellishly virtuous in keeping their promises.

But Satan cannot really create. He can only make with what God has created—make a world with matter and its potentiality of law and beauty. To make what can be tempted and won to evil, the devils must make man with acquaintance: the ability to see beauty; with the function of knowing: the ability to wonder, to learn, to understand, to see the true and the good; with freedom, which may be analytically implicit in perceiving and knowing: freedom to choose wrong and so also the freedom to choose right. All this possibility and actual doing God saw and sees, allows. The devils—especially Satan and Belial and Mammon—have done a right good job. (In doing a good job at evil do the devils get some betterment by the goodness of the job or get worse by the evil of their aim?) God is not sure. He smiles, sadly and hopefully, and lets it go on as he did at first. So the world is a waste which never quite succeeds in not conserving and growing; a folly which never shakes off insight and the getting of wisdom; cupidity that finds itself generous; cruelty making for mercy; cowardice turning up heroes; lust that cannot forget love; and decay that becomes glory. And God can always destroy.

2

Thinking Ways

WHEN WE THINK about thinking, as when we think about remembering and imagining, we are apt to make the objects too much like those of ordinary physical perception. We do this partly by making those physically perceived objects too much like those surrogate ones we think of as the objects of memory, imagination, and thinking. That is, the conventional modern account, the propriety in the matter, is to regard, as John Locke taught us to do, "whatsoever is the object of the understanding when a man thinks" as an "idea"—a mental picture or quasi or pseudo picture. All these ideas are ontologically like each other and each is numerically and (except for an idea of an idea) ontologically different from the real, if any, for which it stands. The "concept" differs a bit, to be sure, but mysteriously and not ontologically, and it serves the role of an idea as being that which is present to or in the mind when the mind thinks or when there is thinking.

Then in the more popular view there are different

faculties—sensing, remembering, imagining, thinking—
to do their respective jobs on the ideas that are there.
With the technical philosophers the faculties are out, as
is the mind itself, and the pictures, the appearances and
concepts, must differ on their faces (which is all they
have) so as to announce themselves not only as chairs
and not tables but also as sensed or remembered or
imagined chairs. Indeed, they must not only announce
themselves (for there is nothing to announce themselves
to), they must so differ in themselves or their aggregation as to constitute memory as different from sensing,
imagination as different from memory, thinking (possibly) as different from the others.

I have liked to ask doctoral candidates, especially in
history, in their oral examinations how they know the
past. They say by present images. Then how do they
tell the present image of the past from the present image
of the present? They say it is less vivid (especially if
they have read Hume). But how do they tell the present
image of the past from the present imaginative image of
the distant in space or the never was, and how do they
tell what part of the past the present image applies to?
This they think unfair.

And it is. It is not cricket to be peremptory with
theories of perceiving and thinking, since none has ever
been offered us, so far as I have heard, which is not inadequate, or absurd, or nonsensical—indeed all three. I
include my own acceptance; although of course I think
it is best.

I would accept knowing as the original fact—mind,

whether it be spirit or matter or organization or what, knows—a fact not to be accounted for or unpacked or analyzed but to be used as barely as may be in accounting for other, less basic, facts. It has been the effort to account for knowing in terms of more complex and artificial facts which has kept us confused and bemused. Further, it is the mind, the self, an enduring and changing thing, a conspective thing, that knows; neither the things known nor their pictures ever are or can be knowing ("knowledge" is a word well avoided). When I use my external sense organs, I perceive physically. When I turn my mind to what is not physically before me and perhaps never physically was, I may be said to imagine. When it is my own past I know, I remember. I actually see the star that was a million years ago and I remember what I ate yesterday; in both cases without image.

I should prefer to keep content firm—and external and direct—in sense perception; to make memory and imagination a matter of "thinking that" we perceive, without any image content; to allow thinking no content except that of perception, memory, imagination, abstractions from them (not concepts), and symbols. I would assign to thinking the processes of association of words or ideas, calculation by rule and ordinarily with symbol, the imagining of things as they are not and the choice of outcome, perhaps with some of the imaginings being of what is incapable of reduction to any items of "experience," incapable of any ostension;

and with the choice of outcome having the distinctive flavor we call assertion, the making of a cognitive bet. The processes of these sorts of thinking are very much the same, but not just the same, as those that surround the central fact of knowing in sense perception and in memory and fancy: direction of attention, entertainment, recognition, discrimination, selective attention, valuational choice. I should like to keep the "faculties," the ways of knowing, as nearly one as may be and find the differences in what is known.

Thinking, surely, is more than the knowing that is remembering and imagining (and sensing). It builds more complex processes around it—unless it is just "thinking of"—for it is "thinking about" and it seeks to add that assertion which Descartes called the intrusion of free will into the cognitive.

And what is known is whatever it is, not knowledge. This is to say there is no mental content, although there are minds and mindings. There are things and their characters, their qualities and doings and relations and values; these are minded directly in perception and indirectly in memory and fancy. Thinking is at least in part and sometimes (or it may be always) made up of these, especially of fancy as a thinking of things or characters as they are not but might be. Yet it is hard to believe that thinking is nothing but this. We come to feel the need to add to our inventory of things and their characters when we try to be clear as to that which we are *thinking of* when we are *thinking about* something. But what we need to add is by the same token just as unclear. When the mathematician talks of the function of a complex variable, when the moralist says it is always wrong to lie, what is it that he is directly or indirectly minding?—what thing, what quality, what action, what relation that could be shown or described or could be denoted by even a perfect phenomenological language? Of course sometimes there may or must be such items. The moralist may picture a little boy with mouth circled with jam and saying, "I didn't get into the jam." The mathematician may use as an instance an elliptic curve—if I understood one of my mathematician students yesterday. Talking of a very pretty discovery of identity and zero he drew such a curve on the board, and I have seen the jam boy in comics: clear ostensions. But this sort of thing is not what we mean or not all we mean by the theory of complex variables or by the categorical imperative.

I think what we need is even more of that separateness of the mind from its objects and its ability to survey and stand off or above which allows us to deal with the differences of remembering and imagining. For even if I do not believe in appearances of things, surely things —present, past, absent,—appear differently to me. Yet the remembered and the imagined bear little firm on-their-face difference. I know, however, what I am doing with my attention—that I am peering into my past, or into recorded past, or into the actual world not now visible to me, into the world of fancy, of fiction. And so, too, I am aware of what I am thinking about even if not of just what I am thinking of—the field within which I have to imagine, construct, choose, assert.

Sometimes, but only sometimes, this is simply described. I have a maze: I am to imagine a path out. I have scrambled letters: I am to rearrange them into a word. I have the problem of perception: I am to imagine and accept that story of things, activities, faculties, objects, assertions which has the most adequacy and the least nonsense and absurdity. But exactly what is before the mind when it thinks of "activity" in order to do its job of thinking about thinking? I have the problem of what is common among the good solutions of my problem, the good steak, the good act of heroism, the good woman. Now the problem seems to founder on its own topic. What is before the mind when we say "good"? It is not hard to imagine something of the several cases and of our "pleasure" or "approval," and then to say it is the presence of the latter we mean by "good": "for we do not strive for, wish for, long for, or desire anything because we deem it to be a good; but we deem a thing to be good because we strive for it, wish for it, long for it, or desire it."[1] But the "feelings" in the several cases are really quite different, and for common sense it seems clear we feel the feeling, if we do, because we see something as good, we do not see it as good because we have a completely unaccountable feeling.

[1] Spinoza: *Ethics*, III, 9.

If there are, as in some sense there must be, some goings on corresponding with what we call present perception, memory, imagination, then it is natural to suppose that some perception is of actually present existing things. (It may be that these are in part perceived as they are, that we can have signs of when they are so perceived, and that they are perceived without the interposition of any idea, image, representation. I hold all three theses.) It has of late been fashionable to become doubtful of the givenness of perception, to let interpretation intrude upon the "datum," to become doubtful of the actual existence as perceived of the things presently perceived. I think the motive of this is quite sound; but often it seems a product of our sudden predilection for staring at pictures of empty polyhedra, stairs, and little creatures alternatively rabbits with ears or gazelles with open mouths.

The older habit was to take present perception as firm content but as representative, and then to take everything from there on as less satisfactory content (or occasionally as at times in Plato and Descartes, to reverse this). "Concepts," which sprang out of, but are probably not in all fairness to be blamed on, Abelard's ingenious contrivances, then became the content for thinking. I have for a long time not been content to believe in concepts or been able to understand what they are supposed to be or to be like, although I have thought I understood what purposes they were supposed to serve. And since early graduate-student days

I have been unwilling to say I believe that representative
ideas make up present perception. I think that when
we remember or imagine we think that we are presently
perceiving (knowing the while that we are not) the
thing remembered or imagined. When we think (intellectually, abstractly, logically, discursively, creatively)
we think of and with present, remembered, imagined
things, qualities, actions, relations (intracorporeal as
well as external), and values. And whatever in all this
seems to be present—the sun on the tent I lived in as
a boy in summer—is so, and not by any kind of representative. But this leaves gaps and puzzles. At least, I
remain insistent on keeping real, direct, honest perception of things and ostensive or some "real" definition
in common sense, poetry, and religion, regardless of
rabbit-gazelles, and regardless of white whales.

And what then is the mind? It is a thing, or a thing-
part of a thing, or a character of a thing as its activity or
set of relations. And the thing of which mind is a part
or activity may be the body, or the soul, or the person.
At any rate mind is not composed of its own "mental
contents"—one of the chief modern delusions, it seems
to me.

Whitehead, Cassirer, Mrs. Susanne Langer have emphasized symbols and the use of symbols—"Man is a
symbol-using animal." Others, like Mr. Paul Tillich,
with somewhat other meanings for "symbol," have
joined the chorus. There is a later chapter on symbols.
Here it seems safe to say that surely symbols are often

what is before the mind or are the vehicle by which the mind does what it does when it is said to be thinking. But the symbol is not original and ought not to be final; it stands for, is vicarious, surrogate (but is not representative except as explicit picture or ikon). Symbols are a way of making calculation possible. As the symbol is the substitute for the real thing so calculation is the substitute for real thinking. Not only may that which is calculated be thought (though more slowly and hazardously than by a good calculator and a good calculus) but also the calculus itself presupposes real thinking sometime before. One may formulate a formula for calculatingly creating a calculus; but this is not original, is not the real creation. Logic lets us avoid the labor of thinking, but sometimes logicians must think.

Now I am not sure whether animals use symbols. They certainly use signs. But if they use symbols in the way of a calculus they must also have some power of real thinking. If man is the symbol-using animal it must be that he is also the symbol-making animal, the animal that can understand and direct and question and deny and prefer. This perhaps is what is meant by the old-fashioned, "Man is a rational animal."

The important difference is that calculation is a matter of rule not necessarily demonstrative and that thinking—reason—is free. This goes along with the fact that calculation is easy and thinking is hard. We seldom take the trouble to be free. Freedom and rationality is an old and a beautiful partnership in human notions. But only too often it has fallen into a calculated pattern of the obedience to law, and the obedience even to a perceived-as-so or a "self-imposed" or "autonomous" law has too much the feeling of the schoolboy following the algebraic formula, and freedom becomes no more than the absence of compulsion by force. Reason is radically free because, although it is perceptive of what is, what is includes value as well as fact; and reason is perceptive of what-is-not in the sense that imagination is; so that the soul that exercises its rationality may, must, create and choose as well as just perceive and follow. Reason is radically unpredictable; but not random as is chance. So nothing is more trustable than reason, despite its unpredictableness. God is free and "in Him is no evil at all," and "before the hills in order stood," "eternal are thy mercies, Lord," "who was and is and is to be," and "our defense is sure," "a refuge never failing"; yet not God himself, I believe, can foreknow or predict (although He can promise and fulfill) what He will do. Our unpredictableness, thus, is not only by the fallibility of our reason but by its essence.

This is not meant to be mysterious. Partly, I should have to admit, it is not understood. But I think it will seem mysterious to some because of the feeling that in every situation it must be the case that there is one proceeding or choice which is the best; that this is a matter just of descriptive characters of what is, or of the possible choices and their outcomes, which could be known;

and that the goodness, betterness, bestness must be a function of some criteria which are commensurable and possess the transitivity of a true serial order. Then it is felt that insofar as reason sees, it must do, as a billiard ball supposedly obeys the composition of forces upon it. I have come to think all these propositions neither needful nor true.

Reason is the cognitive aspect of that aspiration which is the opposite of weariness and both the ground and opponent of depravity. When the mere possible rock becomes an existent rock, it finds it must obey the law of gravity and the second law of thermodynamics (although even here I believe there is that chance which is neither cause nor freedom). If it rejoices in this activity of motion and impact, also it must often be tired and wish for the absolute rest of the merely possible. When the inorganic becomes organic, the plant now finds it must both obey the simple laws of physics and at times rise superior to them; must push up against gravity to the sunlight and reverse the law of entropy by using its chlorophyll to concentrate energy to higher forms. The lower laws become its pleasure but not its glory, and there is still the weariness that calls for cessation. The animal carries on the tale of obedience, of gratification in the carried-over lower nature, of glory in rebellion and superiority, of aspiration in the new movement and sense, and of weariness. And so does man, with a weariness that daily leans back to the animal and sometimes when even the plant life becomes a burden all the way back to the rock and the merely possible; man, with his aspiration now in reason, a new dimension of freedom; and perhaps a new power to use what is remnant from the lower as expressive of the aspirant. The farthest back seems too universal perhaps for most of us to use; but the nearer remote of animal and plant nature become the capital of modesty, the resource both of the indulged exertion of lust and the lyric sharing of love.

"Oft was I weary when I worked at thee." But there is reason to suppose that the pulling of the oar put the boat to some worthwhile change of place. Weariness is not just opponent of aspiration. It makes rest sweet after decent effort, sweeter after some accomplishment. Someone said the saddest lines in English poetry are those about Michael: "Many and many a day he thither went/And never lifted up a single stone." Here is the wiping out of aspiration by the defeat of oneself in another—the defeat of the old man in the disgrace of his son—the loss of the capacity for action which will make weary.

What made me think of Michael? Why should I suppose something "made me" do so? Well, was thinking of Michael pure chance or pure freedom or some complex of just these two? These are my two rubrics beyond the making of something happen that we call cause. As just coming to one, such a quotation cannot be choice, free or calculated. I suppose one could say a sentence, then decide to add a quotation, then think up or look up and inventory a lot of apposite ones, and choose. This would involve a lot of choice but not choice only; and, although I like to quote, I do not believe I have ever done it this way. And equally the thinking of Michael was not just chance, for the lines are apposite, to my attention and my intention—at least, to my preceding words. *That* I should think in this fashion was, I take it, largely chance; *what* I thought of was, we say, "by association"—"association of ideas" we used to say, now not so often. Perhaps it is "free association," but how free is free association? If we go back to Hobbes and Locke we see it is meant as free of deliberative guidance, "unguided" as opposed to "regulated" Hobbes says. I should say, free of free choice and calculation, but should also willingly say, more in their words, not directed, not steered, not led by open attention to the end of the discourse or to the rules of calculation.

It is well to note that things like "thinking of Michael" are in Hobbes's story rather than Locke's. Hobbes's "unguided thinking" and Locke's "association of ideas," though they are taken together in histories and were in history pretty certainly connected, are two. Hobbes's rubric will cover Locke's, which is also unguided; but Hobbes is considering a succession of ideas somehow connected in meaning. Locke has shifted attention to what thereafter was and is usually suggested by "association of ideas": a "degree of madness" in which our minds bring up ideas at the same time which have pre-

viously come to us together in some repetition or vivid-
ness but which in themselves have no "agreement or
repugnancy," are "loose and unconnected" (in Hume's
phrase). So Hobbes's linking is more logical or phenomenological, Locke's more of psychological or indeed
physical cause. Both are unguided by conscious intention. Thinking of Michael is like Hobbes's "malicious
question" as to the value of the Roman penny when the
conversation was on the English Civil War. (Delivering
up King Charles—Jesus—thirty pieces of silver—Roman penny.) It is much less like Locke's "pain and dis-
pleasure" belonging to the pleasant room where one
has suffered or the need of the gentleman dancer for a
piece of furniture like that beside which he had learned
to dance.[2]

But notice how close yet far we can come. To one
who knows the poem the word "Michael" may bring
the idea of, or actual, sadness. There is no essential
connection of the sound, the shape, the word, or the
name "Michael" with sadness. But the story of this
man named Michael is in itself sad. By chance or cause
or by Lockean association our minds may be brought to
the token "Michael," this "makes us think" of the
conclusion of Wordsworth's poem, which we believe
and feel to be sad, and so comes sadness or some further
associate of sadness; all without regulation.

With Hobbes and Locke it doubtless seemed the puz-

[2] *Leviathan*, Pt. I, Ch. 3. *Essay Concerning Human Understanding*, Bk. II, Ch. 33.

zle was to see how people come to connect those "loose and unconnected" ideas. We have come to think we understand that puzzle, for as things or events, which is what our science wants to deal with, they are not loose and unconnected but neurally and temporally conjoined. And even as "ideas" their lack of logical connection is no bar to their being connected in our experience under the principles set forth by Hume (following Aristotle) of resemblance, contiguity, cause and effect. Actually in Hume (with cause disintegrated into association and resemblance played down) these three principles become the one principle of contiguity, compresence in time and space, thus letting us think the ideas have come into connection mechanically and so they reappear together mechanically. So the difficulty comes to be how ideas, no matter how logically connected (logic being no matter), precipitate one another in the actual flow of what comes before our attention. How is it that of two thoughts, as Hobbes calls them, which may never have come to me at the same time, one should now follow the other as though produced by it?

This is easier if it be a case of conscious derivation of one from the other by rule, as in deduction or in much of what is called induction—calculation either in logic or in mathematics. That there is a difficulty is clearer when there is suggestion, recognition, some leap from the one to the other—and it is the leap that seems typical of all real thinking. Even this seems, perhaps deceptively, not preposterous to an old-fashioned believer, as I am, in a real self, which by that power I call reason can see several things, in fact or value, at the same time and imagine others partly like and different, judge, and stand back and judge the judgment. And I suppose if I find this ability of the mind or the person (not of the ideas—no "mental chemistry") not preposterous, I can allow it to operate without conscious regulation or effort or searching, a sort of freedom allowing itself to operate by habit. And there is wide range from our sheerest and broken reveries when half asleep to the very wakeful and alert state of the lecturer who is fully attentive to what he is saying and to his audience, and yet to whom sometimes the continuing discourse seems almost to be carrying itself.

We have worked from a fragmentary thinking about Michael to what it was a fragment of: discourse and discourse freshly carried on, being made not reproduced. I have been using three rubrics: "words carrying themselves along," calculation, and thinking. The last I had associated closely with free choice, not making them identical. Choice is one of my three fundamental rubrics for the characterization of event: chance, cause, choice. Cause and choice "produce" the event; to call it chance says it just happens, is unproduced. But discourse is a series of events and must be explicable under the three eventual rubrics. So far as thinking is freedom and choice it may be said to produce what is said (with the assistance of the larynx and such). The others (words carrying themselves and calculation) and thinking in part, are not true producers but represent our way of distinguishing different modes and procedures of discourse. But as the discourses do eventuate, the modes should be analyzable by the sorts of chance, cause, and choice that enter in.

I hurry to add that I do not suppose these rubrics, these divisions, even the basic ones, to be given in nature and imposed upon our honesty. Such seem always in some degree capable of being differently divided out; the hands can be put back in the deck, reshuffled, and new hands dealt. Chance, cause, and choice served Epicurus (see the end of his letter to Menoeceus) and some others and they have served me so well I should hardly change the list. But I know that, even apart from questions of fact—the mighty debates about "fixed fate, free will, foreknowledge," and simple chance which sometimes fatalists and mechanists and freedomists, Calvinists and providentialists and Arminians all seem to dislike more than they dislike one another—I know that divisions into different sets of powers, faculties, and processes are possible. If, for example, we shake cause loose so that, as sometimes seems in Aristotle with the refractoriness of matter, it is never dependably precise, we could telescope chance into cause. But there are advantages in the notion of cause as precise, apart from the love of the precisionist; and if we were to use chance to liberalize cause itself, we would then still need the intrusion of some chance into freedom, which would mess up the simplicity of our analysis.

Plato says to carve the turkey at the joints. But south of Henry VIII's table we usually carve more than at the joints, and some persons like their slices thin and some thick and some (carvers) like hunks. Anyway the turkeys of science and philosophy have hard-to-find joints or too-easily-imagined joints and much meat. Certainly that fowl that is philosophic psychology, in which we are now indulging, is indefinitely jointable.

Having rejected mere chance, full choice, and careful calculation as the nothing-but way of "thinking about Michael" or in general of the discourse that seems to carry itself along, but having allowed each some role, we still seem to need some help from cause, as a before-to-after precipitation. But cause here is a curiously learned and artificial regimen. Although like mechanical cause in that it works from behind forward, it is unlike in that we trace it in a logical series, not a physical one (it, of course, doubtless has basis in a physical-cause series in the brain cells or neuronic paths and energies). Such cause is also in calculation, in progress by rule as in mathematical and logical derivation: adding a column of figures, going on with a syllogistic or symbolic deduction. Here, too, as in the run-on of association, rational choice is minimized at the time and is further minimized as the recognition of rule and application of rule become more automatic in habit. But cause could not thus work here except for the fact that thinking and choice had earlier been at work, for this cause is by stipulation and acceptance. Here, as not in association, chance, too, is minimized. So the calculation can be built into a machine and the cause made cleanly physical. In mental calculation chance interferes, and thinking steps in to correct, criticize, create, direct; but so too with the computer the expert may step in to repair, ask questions, add variables or operators, to design better questions or a better machine.

The second of our questions about Michael as well as the first included a widely causal, a consequential, word. "What *made* me think of Michael? Why *should* I suppose something made me do so?" That "should," taken along with the "make," could be a clue to the old maze of determinism and indeterminism; of my chance, cause, and choice.

So in discourse which is at all intentional, when we are at all "thinking about what we are saying," notably in "thinking on one's feet," there are three modes: letting the words run, proceeding by rule, really thinking by exercising the inspection and imagination of the reason directly on whatever it is one is thinking about —not on the semantic words or on the syntactic or pragmatic rules. Probably none of these is ever pure, or absent.

There is, of course, also discourse which may be indistinguishable in the hearing from what we have been talking about but which is not at the time intentional: something read, memorized, repeated in hypnotism, produced or reproduced by a machine. Here cause rules, although imposed on materials or processes more than causal. A machine can be made to read from a printed page. Human reading is almost necessarily more than reading. How curious indeed and complex is "reading a paper": we see shapes, translate them into sounds, recognize the generally synonymous meaning of the two, and know the meaning; we applaud, wince, criticize, correct, reject, alter by the use of calculative rule and esthetic judgment and thinking, and chance steps in to make us skip, add, misread.

Where in these modes do proprieties function? Not in cause, which is sub-artifice. Not in reason, which may be the parent of artifice but is not itself artifice but its corrector. Proprieties, like anything, may be the object of reason; and hypocritically among the objects proprieties may be the solicitor of reason; but by definition of reason proprieties cannot be the pusher and causer of reason. Calculation may get its rules by artifice and, even if it finds its basic rules in uttermost reality, must select and arrange them or their derivatives by artifice. But, once given, the rules run the show and are a safeguard from fashion. Nevertheless, it is by fashion some of the materials are put into the calculation. These are calculated with and are not here being criticized (calculation may elsewhere be used in criticism of these same proprieties); so here proprieties may determine as the "material cause" determines. And in associative discourse proprieties are a constant and major determinant in this material way; and, along with habituated rules of logic and rhetoric, in the directive way of an undeliberate calculation; and, most of all, as the solicitor and precipitant of undeliberate choice. Thus fashion set the problems and many of the "facts" of the *Malleus Maleficarum*, the *Hammer of the Witches*. Most of those problems and facts seem to us utterly nonexistent; and our proprieties push us or pull us to ignore or dismiss the book—unless we chance or choose to examine it and find the subtlety and power of its logic and its perception within its context. Then we may find ourselves its admirers or choose to be its admirers—and still be its contemners, remembering Johannes Kepler, or less its contemners, remembering Gilles de Rais.

The end frequently takes us to the beginning. "Michael" begins, "If from the public way you turn your steps/Up the tumultous brook of Green-Head Ghyll,/. . . No habitation can be seen." If you come into the Hopkins campus by one entrance you can turn, a few feet from busy Charles Street, off the pavement by a dirt road up a steep hill a hundred yards or so through big oak and beech trees. Part way up, in the summer time, nothing but the woods—and woods-earth and sky—can be seen; although you know near by are the Power House and Whitehead Hall and the Barn and the tall apartment houses across Charles Street. Just a remnant clump of trees, but it is a true woods. I love the woods, and I love trees better yet; sometimes I think I love oak trees more than anything else in the world. There must be something here of that difference of levels, of the feeling for the earlier, simpler, more basic; for the animal, even the botanical itself; which allows me to feel—which makes me feel—for the trees and allows me to use them in the restful relapse from— as well as in the expression of—the human ideal. For simply on the "higher" present level the statement about my loving oak trees is silly—silly to weigh the love of oak trees in the same balance with the love of certain human beings or even certain undertakings.

That it should be especially oaks among the trees is biographical, of my boyhood summers at camp in a large grove of the biggest white and black oaks I have ever seen—even, if you will, of previous biog- raphies of mine, but not earlier ontologies or racial levels. Now, in objective ways, the white-barked beech trees on our campus are, I think, more lovely than the oaks; but I love the oaks more. If all this, "Be but a vain belief, yet, oh! how oft-/. . . when the fretful stir/Unprofitable, and the fever of the world/Have hung upon the beatings of my heart-/How oft, in spirit, have I turned to thee,/. . . How often has my spirit turned to thee!," and my body also.

Trees, and oak trees: the thorough-bass of prerational motives and of childhood's building, these, although not prior to consciousness as physics may be taken to be, are prior to artifice and so, in essence, prior to pro- prieties. And they very powerfully move our feelings and our projects and our recoveries. Upon and around and against these we impose calculation and design and planning. I believe planning, which occupies so much

55

of some persons' time and so many archival pigeonholes, is pretty futile even in today's doing. The things called plans for a building or a bridge are designs, needed and, if competent, determinative. The actual builder may design his working force and looking ahead to day-by-day exigencies may plan but, if he is a good builder, I suspect he saves that time and meets the day-by-day with sagacity and ingenuity and, of course, Aristotelian good habits. Also upon our background natures and in, through, and around our calculations and designs and plans, are associations, fashion, proprieties—or our reactions against these. And then there is real thinking, freedom, the crown of the rational, perception and creation in fact, in truth, in value.

An example of this "wonder of beauty," real thinking? Here is the fifth race at Pimlico, and I want to make a bet. Probably I will not think much or at all about it. Maybe I have heard "information" and accepted it (why I accepted it is now in the past). I follow the tip, meanwhile anticipating the cash and good feeling I am about to have. I may follow a hunch —like association—some verbal trick with a horse's name or a feeling not based on any consideration of causal factors about to come into play on the track. Or I may calculate: I am a figure handicapper and add, subtract, weigh, and derive figures assigned to the Pimlico track and its conditions. I consider the weather today and each horse's fondness or the opposite for Pimlico, slow track, and muggy atmosphere; the "index number" representing how long it has been since the horse's last race and his last time in the money; the distance of the race and the horse's time for that distance; his assigned weight, post position, rider, trainer, owner and any immediate change of ownership or trainer or rider—and I get a resultant figure for each horse and am supposed to play the highest or lowest.

But I may really think about the race to come. I take all the facts the calculator uses and more (although some I may dismiss) but instead of turning them into numbers or even words I take them in their own right. I assemble them and see them coming together, conflicting, developing, emerging, perhaps from the horse's last year at this time or the previous year (for many horses are seasonal): from this morning's blowout and his record of works; from the morning breakfast in the trainers' coffee room, into the paddock, onto and around the track, the start and the turns, to the climax at the finish line. I know all these imaginings are tentative, but they are thought and not dreamed or fancied, and so I feel the relative probabilities without numbering them; and all this I do while standing in the paddock and looking at the coat and eye and gait of each horse; the grooms, trainers, riders; the saddling and instructions which I cannot hear. The final decision of my judging reason with included scraps of calculation and association sometimes surprises me. And, of course, I also know the most preposterous outcomes are not infrequently actual. All my thinking may fail to turn up a good bet (there are still the odds to be considered) or after the thinking I may even think about the thinking, and in either case maybe I put my money back in my pocket and enjoy the spectacle.

I am convinced I could know perfectly all the facts, of all sorts, and still go wrong. Chance is always there. I used to want to write a story about the man who comes home forlorn after a day of missed long shots and is granted to go back in time again, with the memory of how things have already happened, to the start of that day's racing with conditions precisely the same (except his memory): same horses, weather and jockeys. He bets the winners that were, but a new set wins. This not only could happen (on the impossible and perhaps in part unimaginable assumption), I believe, but in some degree would happen. So what I have said about thinking at the race track, a sort of thinking which may be one ingredient or aspect of planning, does not contradict my depreciation of planning. And the world is larger and less isolated than the race track; large, intricate, and indefinitely subtle as that lesser world is.

The man at the end of the otherwise empty bench down beyond the grandstand at the Laurel race track climbed up on the bench, as the horses came around the stretch turn and went on by toward the finish line to the right. He was very much excited, but very quietly.

"I think he lost and I lost," he was saying—began saying at the eighth pole.

It was a close finish between two horses with the one on the outside apparently winning.

"Who was that out there?" he asked. The photograph sign had gone up. "Number 2," replied someone in front. "Number 2," said the concerned man, "Sir Sag. Sir Sag—that was my top horse." And he took a yellow card from his pocket and looked at it forlornly. "That's good for the card. That's good." But he said it forlornly.

I recognized him as a maker and seller of tips.

"I put all I had left, all my money, on Handle," he remarked to no one.

I knew why. Handle was an even-money favorite.

And Handle was the other horse in the picture. But the man did not know that.

"Yes, Number 2 first and Number 6 second," said front seat.

"Number 6?" the man questioned, "Number 6? That's Handle. If he was second he saved me. I put it all on him to place."

"Of course, he's second. Anybody could see that," front seat said.

"Four dollars," the man said. "My last four."

Then the numbers went up and Handle had won.

And I thought of the old-timer—for this was an old-timer, a tip seller and race follower, not a tout. The crowds say the old-timer does not get excited any more —but this was the most excited man I saw all day.

At the last race, as the horses came down past the crowd, up in the grandstand a youth leaped up and down, waved both arms over his head, and shrieked, "Hit him! Hit him! Hit him hard!" "You may be betting my horse, mister," I said beside him, "but I hope you lose." He did not hear me—and he was not really concerned anyway. My old friend with the four dollars was concerned so much so he could not see—though he knew how to see—that the horse he bet and the horse he picked and expected to bet were out front all by themselves.

He picked one when it came to his last four and played another. Well, he was an expert, he knew enough to know, or feel, how little his expertness was worth. Maybe, maybe not, he also believed his expertness was still the best thing he had to go on. He was not asking people's quarters dishonestly. And yet for his last four, he could not resist that even money.

And I thought of Descartes.

The next race I was sitting down on the front bench when the horses came on the track. There was another even-money favorite, though he had not opened as such. And a man I had never seen before came, stood behind me, and observed, "They sure went to that Number 1 —bet him like they knew something." Then in what seemed to me utter contradiction, "Yeah, he looks like a cinch. I don't think they'll ever be close to him." To a friend walking up he called out, "Well, any excuses? Find any way the favorite can lose?"

This favorite was not close, but he did get second.

And why did I think again of Descartes? Because

last year I had been struggling to find out what Descartes meant in his central emphasis upon knowing what we mean, upon clarity, upon "dividing difficulties"; and one of the things I decided he meant was a call for the actual imagining of the actual operation of whatever it is we are talking about. And it can be added that it is often hard, sometimes impossible, to do so competently, and that it is easy, tempting, to skip it, even sometimes in fields where our familiarity should make it easy. Often since then, but especially at the race track, I have been reminded of that.

Favorites are made by people betting. Yet one of those who make a living by advising people how to bet will run from his own advice to follow the bets he supposedly is directing. Is he swayed by the apparent weight of contrary authority? Does he suspect "stable money"? It could be; but he has ways of checking the facts and he has seen such authority and such money lose many times. Last night he would have said "So what?" Usually he does not stop even to say "So what?" but goes on and bets his selections. Today, for his last four, it is just that even money; no hypothetical figuring on what he might imagine, only the massive unimaginable.

And the man behind me: he had figured the entries and Number 1 did not then seem a standout. But after that even-money sign went up and stayed—"He looks like a cinch. They won't get near him."

And those who complain an owner did not tell them

his horse was going to win; do they ever ask themselves precisely how the owner came by the information he failed to impart? And those who talk of fixed races or of an owner's "not letting him win today"; do they ever ask themselves just what they think of as actually being done?

There are ways of fixing a race—not often tried and often failing—but it is not accomplished by a magical decision, even a decision of all the owners or all the trainers or all the jockeys or even of all of these—and of all the horses. It is simpler to keep some one horse from winning than to make some one horse win, but still is not magical, especially after the prices are seen. All that the most accomplished rider can do is to intimate his desires to the horse and encourage him to comply. The horse is carrying the rider.

But I started to write about Descartes and clarity. Is it not strange that the preacher of clarity who is regularly, and not undeservedly, praised for his clarity, is clearly not clear as to what he means by his method, especially by that rule most often repeated as his: the rule of analysis—to divide so as to be clear.

Well, we "divide difficulties," as he tells us in the second of his four famous rules in the *Discourse*, sometimes when the difficulties are practical and temporal by considering successive performances or phases of performance. To do so and to consider and actually to face one performance at a time (even though we also need to have an eye to the others) is excellent advice practically; and it has its advantages for science also, though these are lesser.

And we can divide a geometrical problem this way if it is one of construction and we can do it spatially in the case of some of the theorems. This was, of course, effective in Descartes' thinking.

Then there are the divisions of objects, like chairs, and of assemblages, like clubs. We can divide a chair into its structural parts: seat, back, legs, rockers; and into its material parts: blocks, chips, and splinters of wood, perhaps molecules, atoms, electrons. And we can divide a club into its member clubs or member individuals.

We can also divide "chair" into its sorts: rocking chair, straight chair, arm chair. And we can divide "chair" into its definitional parts: a-piece-of-furniture-for-sitting—into genus and differentia and each of these into theirs if any.

We can divide not only a club but also a nation into its federal states and its citizens, its members. We can divide an animal or vegetable organism into its members, which are neither federal nor individual. We can divide a church, a board, a mathematical set, a logical class, an assemblage, a species into its members.

We can divide a thing into its existence, its quality, its activity, and its relations. We can divide its quality into its component qualities. We can divide color into its particular shades.

We can divide a motion or a force into its components—as Descartes did in his geometrical analysis of force and motion.

These are, at least some of them, very different. Any may be but none must be at any time the sort of composition and division requisite for Descartes' purpose.

One frequent combination of divisional tasks is where we have a "vague" word or a "mixed mode" which must be defined or at any rate have its denotations sifted out and then these isolated in temporal and causal successions. Part of this is Locke—another of the ways in which John Locke is a good Cartesian.

The same requirement is behind many of the vague rules in present textbooks of scientific method, particularly on the subject of hypotheses. "Hypotheses must be clearly formulated," and "formulated so as to be capable of being tested." These chiefly mean "You should know what the hypothesis means in the sense that you can describe the actual occurrences supposed to follow one another in whatever context it is you are working in."

The safest—Descartes was a formalizer of this course —is hypothesis as a story of the motion of simple things in space and time. For here our imaginings are clearest and most communicable. I do not see, however, that one can command this sort of story of hypothesis ahead of time.

And this perhaps was Descartes' trouble (as it had been Bacon's with his other type of motion doctrine). He wanted and he got this master rule of space-mo-

tion in severe form, and it was a rule that worked beautifully for his longed-for mathematical physics and
anatomy and that continued to work amazingly for much
of future natural science. But he did not want just to
announce or prescribe the world as the world on which
his method would work. He was offering a method, a
radical and universal method. He wanted to prove the
material he worked on—at least to prove the justice of
his choice of that material—in his own advising, by the
tests of his method. And the method itself, should it
not be accomplished by the method? At least the explanation and justification of a method fundamentally
commanding clarity should not be clearly unclear. Yet
the material for, or embodiment of, his method material, seems hypothetical or empirical. The method does,
I think, find its justification in intuitive clarity, but
its abstract explanation is far from clear. And it seems
to be hard to make Cartesianly clear either the necessity
of space and motion or the limitation of clarity to space
and motion.

I do not think clarity should be so limited. I think the
Cartesian method is a general one, although not every-
where equally adept. The use of the method is subse-
quent to the acceptance of some subject matter: Cartesian space-time and motion for Cartesian physics;
phonetics and meaning for linguistics; horses and
humans and weather for horse racing. Can the method
take as its subject matter the question of what subject
matter should be accepted for study? Possibly, but only

with some specification of purpose and hence of some genus of subject matter. So it might be said we do not have to be Cartesian in physics—we can be Aristotelian; or use Bacon's qualitative, internal, quasi-chemical motions; or look for a new physics in quantum mechanics. And it may seem that Cartesian physics offers a maximum of Cartesian clarity, not only among rival physics but also among nonrival topics of curiosity.

So the four rules in the *Discourse* are distressingly unclear. Repeated back to teachers by generations of students they make me think of Bacon's phrases about the universities, where "alas, they learn nothing but to believe: first to believe that others know that which they know not; and after, that themselves know that which they know not."[3] And yet they are the rules for clarity by a thinker and writer not undeservedly famous for clarity. So, earlier, the first eleven and a fraction rules of the *Rules for the Direction of the Mind* are involved, abstract, and probably more unclear than the brief passage in the *Discourse*. It is not until after the opening paragraphs of Rule XII (where "Selections" are apt to cut off) that light comes, especially in the examples. And then it comes with apology, for the treatment now has to be based on a psychology-physics which, no matter how neat, Descartes cannot claim as established, but only as hypothetically possible and here serviceable.

Kant might have justified it "transcendentally" as

[3] "Mr. Bacon in Praise of Knowledge."

necessary for the method (and *this* method doubtless derives from Descartes in part) but Descartes had not got around to this; and of course the assurance of the method could be denied and it is still more of a question whether it could be made out that the space-time-motion manifold is the only allowable physics for a Cartesian method.

Later on, with the aid of the cogito, the self and the notion of "attribute," God, and the world, Descartes thought he had clearly established his long-held and long-bothersome hypothesis or faith of extension and motion. But he seems not to have thought so and not to have rearranged the theory of matter into extension until some time after he had told the story of cogito-self-God-world in the *Discourse*. From the *Meditations* on, this helped him to be clearer about the method—after he had quit writing books about it. But the proof is surely disputable.

And it seems curious that the key step—one which seems more than he was entitled to—was taken in the *Discourse* in respect of the self to give it its one attribute of thinking. If it were then applied to the outside world to give it its one attribute of extension his desired establishment of his physical faith would have been made. And in that case the historian would be tempted to say Descartes' need had moved him to the extra taking. And that might somehow be, but at any rate Descartes put off cashing in on his venture. The step is taken in one sentence, after the paragraph of the

"I think, therefore I am" (and the argument of the one sentence is never really added to here or in later works). "From that I knew that I was a substance the whole essence or nature of which is to think . . . and even if body were not, the soul would not cease to be what it is."[4] Descartes might claim that by the cogito he has shown both that he exists and that consciousness, "thinking," is not only the sufficient but also the only ground of such showing. I do not see any Cartesian clarity in the inference from this that thinking must be the only attribute or even that it is an attribute—essential character—of the self. Thus, even if one presupposes (as Descartes does not seem to) that every substance has one and only one attribute, the fact that, *in ratione cognoscendi*, consciousness is the one way of proving one's existence does not say that *in ratione essendi* consciousness is inseparable and the only aspect inseparable from one's existence. Suppose a person is a body which once in a while happens to be conscious. In such moments and only then he would be able to be aware of existing and to prove to himself that he exists. But all the while he goes on existing. I have come to feel that Thomas Hobbes in his "Objections" to the *Meditations*—for all his apparent unwillingness to be fair to Descartes—is right in his pivotal denial of the security of Descartes' proof of the simple thinkingness of the self and that Descartes is even more apparently impatient of understanding Hobbes. This reverses what I once felt.

[4] *Discourse on Method*, Part IV, Haldane, Ross I, p. 101.

And it remains that a method of spreadout point-to-point continuous-line clarity is, as Bacon insisted science should be, a method of "dissection" rather than "abstraction," but that abstraction cannot simply be eliminated. In nature dimensions are more real than sets; and sets, in logic, are less real than classes, and classes less real than characters. An extensive logic always has an advantage: in applicability, power, calculability; and is always inadequate in the end. Every now and then the powerful extensive calculator catches a glimpse of his bareness and cries for help from his confused intensive mother. Extensive logic, too, abstracts; indeed it abstracts the more in that it abstracts from "extension" and sets which have already left out (been abstracted from) much of the concrete world. But thus it gets closer to the map, and to that form of abstraction we call mathematics; it gets further from words and meanings. When we "divide" in definition and classification (going up and sideways to genus and differentia or down to species) we are abstracting in Bacon's sense. Can we eliminate this or the syntactics that go with it? Surely the worldly, sagacious, inquisitive politician and essayist who "wrote philosophy like a Lord Chancellor" did not eliminate this even in his science. Indeed any "real physics" in Bacon's demand or any metaphysics must take account of *sorts* of things, of the difference and relations of sorts, characters, and things, of some real definitions—must be more than a reproduction, a proper-name inventory, or a Baconian or Cartesian map or geometry thereof.

I have found that students almost unanimously when they want politely to call some notion or statement fuzzy, vague, obscure—a case of not knowing what one is talking about—will use the word "abstract." There are such abstractions, many of them the descendants of Locke's "mixed modes." But abstraction began (at least half of it) in pursuit of precision, simplicity, clarity —Cartesian virtues. Even the abstracting of classes and definitions can achieve a clarity of its own, getting in under Descartes' method although not now of visual, geometrical, or physical dissection.

And there are two other considerations favoring "abstraction." We can sometimes, perhaps always, maneuver up on classification or even metaphysics by putting it into a descriptive story, thus using a sort of Cartesian physics. An English-composition way of stating the Cartesian method is to say that argument or exposition must become description and narration, and that any part of it that is to claim clarity must have no gaps. Descartes himself called on this and gave the *Discourse* some of its readability and understandability by putting much of it in the form of a biographical story.

The other consideration is that Cartesian physics is itself a severe abstraction and in history on through Newton to Einstein becomes more so—not an abstraction of classes but of one aspect of the "dissectible" world, its time-space-motion. That all the actual substantial world can be reduced to and in some sense predicted from that rarefied aspect is the Galilean-Cartesian (and, with a difference, Baconian) assumption: a radical if not a rash one, and it is no wonder they shunned announcing it. This abstraction retreats closer and closer to the master abstraction of mathematics. It is the "machine in the ghost" of the farmer's real world of red, juicy apples and of God that stirred Berkeley to revolt. It becomes true of more and more, and less and less true of the more and more; or, if you please, more and more true of a more and more severe abstraction, a more recondite substructure, of the familiar world. And it becomes more constructive, more formally stipulational, in its abstraction. Newton's law of the inverse squares for masses taken as ideal points follows Descartes' vortexes; non-Euclidean geometry brings a new provisionalism and helps Einstein find a more submerged formalism to get together Newton's particles and electrical fields. (Quantum mechanics may be closer to Bacon than Descartes.) But physics always keeps hold of the actual world at least through some tie of operational definitions and measurement. It is measurement that holds descriptive geometry to metric geometry, and it is physical measurement—meter, gram, second, and how many—that holds metric geometry and formal physics to the actual world; that bridges the universal-individual gap and the "inductive leap." Not all things, most will say, can thus be measured and so not all things can share the Cartesian method at its best.

At any rate this is one effort to think about thinking, one account of *bon sens* as the intuitive reason or light of nature that can be the occasional overseer of cause and association in discourse, can be the associate of freedom in its adventures in the event world of cause and chance; and so is at once the enabler and the critic of proprieties. It would not do here to deny—and it is not needful here to affirm or deny—that there is a mysticism beyond reason just as there is a faculty of immediacy in reason; nor is there any wish to praise Descartes' as the only or the best methodization of the uses of reason.

All manners of thinking have their uses; but for intellectual knowledge certainly and perhaps in some sense for any, even mystic knowledge, one has to look back to and forward to some Cartesian ideal of completeness and clarity, with nothing left out that is relevant and nothing left in that is unresolved, so that intuition, induction, and deduction (the three modes of the *Rules for the Direction of the Mind*) may freely inspect, survey, infer. But also we learn—in a sense something becomes clear to us—by a sort of synthesis not Cartesian, by dint of doing; as we may say that in the three hundred years since Descartes it has become clear to us that clarity and thinking are a longer task, have more to do, than Descartes and Bacon thought. I think we have learned that assurance is more apt to be negative than affirmative (compare "These two lengths are equal" and "These two lengths are not equal"); that being sure that we are unsure whether two, especially quantities, are the same (consider "I think these two lengths are equal but I do not know") is the birth of the invention of instruments and so in part the birth of science. I think we have also learned that certainty is in relations rather than in qualities, but that in relations is also the optional and the arbitrary (we may argue about the name of a color but we have to see it as we see it, whereas we may see, and describe, geometrical steps going this way or that way); that among the clearest "that's" of our awareness is the awareness that what we perceive is vague, confused, not clear: the certainly (but not unchangeably) uncertain, the clearly (but not hopelessly) unclear.

3

Symbols[*]

THE MORE ACTUAL for the more real: if there is any rule for symbols this seems to be it—but roughly. So the familiar is used as symbol for the unfamiliar, the clearer for the more obscure, the small for the large, the more trivial for the important, the easier for the harder, "the less primitive component of experience for the more primitive" according to Whitehead.[1] All these comparatives figure, of course, as in the estimation of the user, at least of the originator; for symbolizing is deliberately or implicitly intentional. So the written shape or the uttered sound is symbol for the word, the word for what we want to say, the swastika for lightning and for Hitler, the fox for cunning, Griselda for patience, Dante's trip for the pathway of the soul, rosemary for remembrance.

* To the Johns Hopkins Tudor and Stuart Club, May 15, 1959. The introductory paragraphs have been revised. A part of the conclusion, because of time, was not read.
[1] A. N. Whitehead, *Symbolism, Its Meaning and Effect.* (New York: Macmillan, 1927), p. 10.

These represent many sorts, by no means all the sorts, of items which serve somehow as signs and to which the word "symbol" may be, but does not have to be, applied. "Symbol" might be used of some of the things meant by any of the following words: sign, mark, significative, note, word, character, token, indication, index, type, signal, ideogram, hieroglyph, representation, copy, icon, resemblance, recognizance, model, pictogram, pattern, image, instance, example, insigne, emblem, colors, regalia, crest, favor, badge, totem, signature, ensign, sacrament, ensample, figure, trope, metaphor, simile, analogy, allegory, shadow, paradigm.

Three sorts of symbols seem chiefly to attract the assured or eulogistic word "symbol" in current fashion; at least these are chiefly the topic of this paper.

Say "symbol" to a mathematician or a logician, he thinks of the plus sign, the horseshoe, his variables, x, y, a, b, P, Q. These are made with express intention, are arbitrary in the sense that the iconic basis if any is forgotten or unimportant, and are stipulated—often need to be expressly stipulated not only for full formality but also for general understanding since the users of these symbols are well-known for their separatism (logicians fail to agree on a symbol for the negative although most of them agree negation is their least replaceable notion).

Say "symbol" to a literary man, he thinks of Rimbaud, Yeats, Wordsworth's skylark and Keats' urn, Faust and Mephistopheles, and of course, the white whale. These symbols are made, discovered, contrived with more or less conscious intention or are allegedly so, are free in respect of their authors but not free in respect of their objects, where some iconic, analogic, allegoric, naturalness of connection is essential. The question of possible mistake in the critic's allegation of the author's intention is real here but hardly admissible with the preceding class of symbols.

Say "symbol" to a psychologist or sociologist, he thinks of Freud, dreams, an act of retirement as expressing "the" wish to return to the security of the womb, nature gods, "conspicuous waste," and the gray flannel suit. These, too, have necessarily a sort of intentionality but typically one that is unconscious or forgotten. The flavor of allegation in the assertion of these symbols is stronger; the allegation seems almost incapable of testimonial confirmation and unneedful of it. We can ask a poet whether he meant to do so and so, and most of us would give some credence to his answer. Here we are rather in the position of the interpreter who is more interested in the poet than the poem. Milton, the interpreter may say, used that "fairest of her daughters, Eve," as symbol of a proper woman's unselfish devotion; that he did so was symbol of male incompleteness and Milton's frustration with his first wife. Since these psychological-sociological-anthropological symbols are meant as after-the-event explanations and as revelatory, the assertor remains assertoric despite any conscious denial on the part of the supposed user of the symbol.

And, though this is unfair to psychology and sociology, when a propagandist picks up the technique, the assertion often seems the most confident when the assertor has the least sympathy or acquaintance with that which he is accounting for. Some of the world's most logical and cogent analyses (as in the *Malleus Maleficarum*) and some of the most ingenious and eloquent explanations have been of things that exist only in the beliefs of the analyst or the explainer.

The world has had its fashions in the extent of its symbol-using and symbol-seeing and in what symbols it has used for what. He who thinks of the love of a child for his father and of the father for his child as a sign of God in the world thinks of God as more real than the actual psychology he finds. He who sees God as symbol for the child's later remembered love for his father, thinks psychological "drives" are more real and acceptable as ultimates than "God"—the intended meaning of the actual words he hears. We would learn about symbols and about history if we had a historical almanac to tell us how much symbolism each age went in for and what symbols it used and what it meant by them. We work without it.

I accept two basic sorts of real: things and their characters. Then there is artifice: what we do with these; for example, language. So there are signs: "natural" (although nothing is a sign except in the recognition of a mind), and "arbitrary," and the symbolizations of the art of living and of fine art, all of which are both natural and made up. "Symbol" and "symbolic" have not always been the word: in Alexandria "allegorical" was fashionable and in the later Middle Ages "analogy." A wider and more recondite word might be safer than the present "symbol"—say "anagoge" and "anagogic" —with appeal to "man's glassy essence," Shakespeare, and C. S. Peirce. But since today's usages are here my target, today's word is better. Perhaps I can put a case using both: I willingly concede that man in his communicating is essentially, although only in part, anagogic; but I have protests, also partial, against today's devotion to the symbol.

Artists who contrive symbols and the critics of their arts will argue, praise, and object to the skill of the contrivances. Psychologists and sociologists and historians of ideas may praise or object to the adequacy or consequences of the asserted symbols. With these we are not concerned. External or philosophic objections may come from someone who thinks the contrivers or the descryers of symbols are too fond of the game. He may also think that because of this they are letting the symbol swallow what is symbolized, forgetting the normal assumption of the greater reality of the latter. Someone may object that from the viewpoint of certain objects they ought not to be treated symbolically or from the viewpoint of certain symbols that they ought not to be used as symbols or that certain asserted symbols are not symbols at all. Someone may object that the ontologic vector of the symbolization is misread or misdirected.

"He heweth him down cedars and taketh the cypress and the oak. . . .

"He burneth part thereof in the fire; with part thereof he eateth flesh; he roasteth roast, and is satisfied: yea he warmeth himself, and saith, Aha, I am warm, I have seen the fire.

"And the residue thereof he maketh a god, even his graven image: he falleth down unto it and worshipeth it and saith, Deliver me; for thou art my god. . . .

"And none considereth in his heart, neither is there knowledge or understanding to say, I have burned part of it in the fire; yea also I have baked bread upon the coals thereof; I have roasted flesh and eaten it: and shall I make the residue an abomination? Shall I fall down before the stock of a tree?"[2]

Hamlet: O, the recorder. Let me see.—To withdraw with you:—why do you go about to recover the wind of me, as if you would drive me into a toil?

Guildenstern: O, my lord, if my duty be too bold, my love is too unmannerly.

Ham. I do not well understand that. Will you play upon this pipe?

Guil. My lord, I cannot.

Ham. I pray you.

Guil. Believe me, I cannot.

Ham. I do beseech you.

Guil. I know no touch of it, my lord.

[2] Isaiah, 44:14-19.

Ham. 'Tis as easy as lying. Govern these ventages with your finger and thumb, give it breath with your mouth, and it will discourse most excellent music. Look you, these are the stops.

Guil. But these cannot I command to any utterance of harmony. I have not the skill.

Ham. Why, look you now, how unworthy a thing you make of me! You would play upon me, you would seem to know my stops, you would pluck out the heart of my mystery, you would sound me from my lowest note to the top of my compass; and there is much music, excellent voice, in this little organ, yet cannot you make it speak. 'Sblood, do you think I am easier to be played on than a pipe? Call me what instrument you will, though you can fret me, you cannot play upon me.[3]

> Oh who can hold a fire in his hand
> By thinking on the frosty Caucasus?
> Or clog the hungry edge of appetite
> By bare imagination of a feast?
> Or wallow naked in December snow
> By thinking on fantastic summer heat?[4]

We are symbols and inhabit symbols.[5]

Let me put the chief point of my thesis, which is a complaint, at once and bluntly; asking you at the same

---

[3] *Hamlet*, Act III, scene 2.
[4] *Richard II*, Act I, scene 3.
[5] Emerson, "Oversoul."

time to believe that I shall add many qualifications and distinctions and that I have more than I shall add. Much symbolic interpretation makes me want to put Hamlet's indignation into the mouth of what is being symbolically interpreted. "What an unworthy thing you would make of me." And this is the more so as what is interpreted is close to what Hamlet is or is thought of as being, a real person, at least a real thing. I am a substantialist: I think the basic inventory of all that is would be at least in part and in irreducible part in terms of individual, existent, continuing and changing things like you and me. Our existence and reality make us poor symbols. Men and women of history become symbols only as they lose their individuality. So, too, I think it is with the gods, prophets, and persons of religious knowledge or faith. Persons created by the artist to be as if existent are, like us, poor symbols. But, as much more than symbols, we and they have a right to some of Hamlet's protest (though most of us poor actuals have de-realized ourselves more than Hamlet did) against being turned to a lesser role we are not fitted for.

Those of you who are explicitly or implicitly by present fashion on the symbolist side, and therefore suspicious of my antisymbolist beginning (and I imagine most of you are), and also those of you who are accustomed to keeping your logical suspicions attentive through at least the first two or three pages of such a paper as this (and all of you are here)—will already have noted several challenges you might make to my use of the text from Hamlet. Here you should say: "I don't see that Hamlet is objecting to being used as a symbol. What are Rosencrantz and Guildenstern making him a symbol of?" And you will be right but not sufficiently so.

I could escape by saying that all I need of Hamlet's speech is that, just as Rosencrantz and Guildenstern are assuming to pluck out the heart of Hamlet's mystery by finding out why he is putting an antic disposition on, so the critic who uses Hamlet merely as a "symbol" of intellectualized indecision, or of noncommitment or of the death-wish-where-the-sexual-drive-is-aristocratically-inhibited is assuming to pluck out that mystery. And other symbolic uses go further and may do worse except that there is a measure of caution and of truth in the very multiplicity of interpretations. But I am not content simply to let it go at that.

There is a sense in which Rosencrantz and Guildenstern *are* making a symbol of Hamlet—they are forgetting his substantiality and putting him into that other great field of being sometimes called abstractions, of characteristics or possible characteristics of substantial persons or things. Or, perhaps, they are putting him in that intermediate and secondary field of signs, symbols, and surrogates for the real; both of which fields share the difference from substantial things in that they can be seen through and cannot be "made an unworthy thing" by being pried into and understood. (There is some paradox here, for in the older usage the symbol is supposed to be the more concrete stand-in for the more abstract. Well, in the older usage the symbol was ordinarily in what I have called the derived field and stood for something in the abstract field: the figure of the cross is the symbol of Christianity, the picture of the lion the symbol of courage; or, perhaps, something in the field of artifice is the symbol—though here the word was apt to be "emblem"—of something in the substantial field, as the figure of a lion may be the mark of an actual king of England, say Richard I. To this I shall return.) At any rate the abstract and the artifice invite theoretical knowledge—may defy it but invite it. The substantial, the person, properly invites love or battle, not explication. Rosencrantz and Guildenstern are making Hamlet an object of study and using him as a counter in their game of preferment with the new king.

This ignoring of substantiality in favor of prying I take to be one of the vices of our age. It has wider bases and ramifications than my topic but let us note the fashionableness of our psychological curiosity. The anatomist, the physiologist, the pathologist, the medical doctor are among our greatest benefactors and they have their platform of safety in the attention to the body with its objective parts; but even they become, of necessity but at some peril, psychosomatists and confessors. The personal psychologist, psychiatrist, psychoanalyst, too, we must have—but where is there safety from intrusion and falsification? And sometimes they do play upon us—sometimes unfret us. And when the doctor passes from the observable parts of the body to the unobservable "parts" of the self, mind, soul (the "personality" it is now, I believe), a whole new pertinence if not necessity of a new set of symbols comes in. What are will, intellect, instinct, id, superego but devices to stand more or less truthfully for doings of the real person, the substance?

Now, of course, Hamlet is not literally substantial. That there may have been a historic Hamlet, Prince of Denmark, in Elsinore is something else again. And it is commonplace that he is more real than most people in that he is more interesting, greater. But he is, as I have said, "taken as being" real, and by this I mean not only that Shakespeare intended to write of him so that the reader would think he was writing as he would write about a real person, but also that Shakespeare himself dealt with him in great part as though he were real. He clearly learned more about this "creation" of his after his first reading of the story, on to his play as represented in the First Quarto, and so to the Folio. Did he decide to put in extra touches in a consciously contrived construct, for dramatic or poetic or even psychological reasons? I don't doubt that he did, but he did more than this. I think that fundamental to such touches was the fact that Shakespeare thought and felt about Hamlet and learned about him as about a real man. He learned about him, that is, not by study or analysis or prying but by living with him; so that at once much became clearer and the mystery became richer and deeper because its "heart" was left in place.

This seems to me a very realistic view and I would distribute it to almost all the great characters of literature. But it does not seem "realistic" in the technical sense of literary schools if I understand that usage. The "realist" restricts himself to observable facts or materialistically acceptable simple motives and puts everything of this sort—especially everything unpleasant—in. Imagine—or rather let us not imagine—Hamlet as written by this sort of a realist. On the other hand the symbolist (in a narrow but not the French sense) might have his own moral, or his own metaphysical or psychological or even political lesson he wants to impress, or a set of emotions and impressions he wants to suggest. He adds to the framework of Hamlet whatever will and only what will serve as analogy or metaphor for that further vision. The actual Hamlet quite breaks through either formula.

I know, however, that there are those here who have much more familiarity and more competence than I have in literary history and in the interpretation of literary imagination and methods. And this part of my doctrine, while an ingredient and perhaps helpful to my other conclusions, may not be logically essential to them.

All that I have said of those conclusions so far may perhaps be summarized as: No chronicle of a real person, or history, or poem or novel or tale or play which makes up a character to be taken as a real person should use that person or those persons simply to symbolize some ulterior interest of the author or the reader, whatever that interest may be. And it seems a sort of corollary of this that, although the "omniscient author" may be tolerated as knowing all that happens to his character, he is not easily to be tolerated as knowing all there is about any person, even the simplest, even his own creation—first because there is no way to know all there is of what is actually there and second because some of what he must presume to know is not there to be known anyway.

I hurry to get in two of my qualifying acknowledgments: that there are many other symbolic modes, and that even here I am not presuming to prohibit. Some of the other symbolic modes I shall consider. I am sure some of you have been saying to yourselves, "Even if Hamlet is not a symbol, Hamlet himself seems to use the recorder as one." The matter of prohibition has two aspects. I have built my case in part on my substantialist metaphysics—the primary reality of the individual thing, not of its stuffs or elements or equations or drives. This metaphysics is so far from being standard in my own craft guild of philosophers—I think it pretty normally is standard outside—that it is not fashionable, although it is becoming more so again. But I certainly cannot impose a conclusion resting on a premise others do not accept. Indeed, here it may be a direct begging of the question since I suppose one basic difference between my feeling and that of the anagogic opposition is in my substantialist wish to hang on to as much as may be of literal fact. On the other hand my argument even as built here is not just on that basis and, even so far as it is, it is, as is every argument, built on one's own major premises and is properly presented if honestly presented as such—for the concurrence of those who also accept those majors, for the hoped-for concurrence in both premises and conclusion of those who have not taken any position as to the majors, and even for the hoped-for concurrence in the conclusion of those who deny the majors since true conclusions are often derived from false premises.

And indeed I can argue in the opposite direction, from the unsatisfactoriness of this mode of symbolizing to the metaphysical priority of the thing.

Still and nevertheless, although I argue for one style as better and against another as partly mistaken and inverted (like a fourth figure syllogism when a first figure is available) I would not prohibit the worse. I have no prerogative of prohibition, and I am a patron of variety who thinks that sometimes a better and a worse are better than two betters of the same sort, and anyway, even if I thought the symbolism in question is worse than I do think it, I should not wish to prohibit it. I think, let us say, the taste for the tyranny of the telephone is worse, but I would not prohibit the telephone —I have friends who scarcely survive the day without that busy way of wasting time. And the world seems full of people and indubitably intelligent and learned ones who would find little interest in the one-legged captain and his white whale except as symbols of pride's war against evil or something, and I should not want to take that from them or to take from Melville his fame at this late date. I should note there may be differences in using the man, the men, the whale, the ship, the voyage, or incidental devices as symbol.

We use the nearer, more familiar, more concrete, more easily handled, more available to symbolize the further away, less familiar, more abstract, more difficult and problematic; also we use the less real, less important, less threatening or promising, less enduring to symbolize the more real and important. These two motives are part of the reason that two ages, peoples, circles, persons can sometimes take the same pairs in opposite directions, and indeed that the same age or person can go both ways.

We talk of nature gods, we read Cassirer and say that for their devotees these gods were symbols of natural processes: the Hopi rain god (was there one?) was a symbol of rain. I suppose for the anthropologist or present semanticist rain and the social need for rain are obviously real and the god obviously unreal. But I suspect that for the Hopi—though I have never talked with or even seen a Hopi—it would be nearer the truth to say rain was seen or is seen maybe as a manifestation, sign, if you please symbol, of the rain god. And common sense would say the notion of explanation and of production and of getting something produced—and quite literally—is also there.

When we see the trash can turned over and say, "That dog has been here," we don't use the dog as a symbol of the trash can's overthrow or of trash-can-overthrowness universalized. And if we say, "Take it to City Hall" we don't mean that City Hall is a symbol of interposition, retribution, or rescue. I think Cassirer and Mrs. Langer are quite right that there surely is much more than utilitarian consideration in early or any religious doctrine and practice. There is ritual for its own sake and there is satisfaction of curiosity, which go on into worship and science. But when explanation, the satisfaction of curiosity, shifted with Thales and the Milesian Greeks, it shifted, without knowing it, in the direction of symbolic statement, not away from it, it gave up, and not only gave up but renounced, the ideal of literalness; but poetry and religion, together with common sense, did not give it up. In so far as they do they become, or are in danger of becoming, science. There is a great glory of science and hereby we increase that glory. But there is one glory of the moon and another glory of the sun; and when everything becomes science and symbolic the glory of realistic vision has departed. One doctrine might properly welcome this, I think, and only one: the thorough mystic who thinks science and history and poetry and religion and all discourse are symbolic but that beyond all these is the unspeakable vision which alone gives reality. I assent to the mystic claim of vision but not to his claim of truth, still less to his claim of a monopoly of truth or of the vision of the real.

Now among the purposes of symbolizing in literature even among writers the least prone to liken themselves to scientists is the pedagogic, the argumentative one of proving some point in the symbolized doctrine, of assuring its acceptance by the persuasiveness of the symbol. The symbol here, to be sure, does not explain as the scientific theory with its symbols explains the symbolized fact or as the scientific model explains the theory— rather the literary symbol explains to the learner (reader or writer himself) *how* the explanation explains, how the theory is to be understood and so wins its acceptance by the clarification the symbolic story lends. (According to Francis Bacon there is also the motive of concealment.) According to Whitehead, for the poet the actual scene is the symbol and stands for and suggests the words, which for the poet are the "meaning" of the scene. This is accurate on Whitehead's definition and theory but they are concerned with a very bare and basic level of symbolic activity. In those symbolisms we are apt to think of in literature the poet presumably in some degree has the "meaning" and the reader in mind and picks out or creates the symbol, the symbolic object or tale, so as to convey, enforce, evidence the meaning to the reader.

This interferes with the fundamental rule of for-its-own-sake for one who accepts it. It also raises the question: why, if the author wants to convey a meaning, doesn't he do so directly, and why if the reader wants to learn meaning—say ethics—doesn't he do so directly? Some, of course, both authors and readers, do; but many do not and are frankly or implicitly impatient or contemptuous of straight theory while avid of the same subject matter delivered in the artistical and symbolic. This situation partly accounts for the wariness of theorists like me who do like to take our theory straight; a preference spiced for the professional with a degree of envy that so many who are amused or bored by the idea of studying the *Nichomachean Ethics* or the third book of Hume's *Treatise* will study and discourse long on moral philosophy as taught by Faust or the white whale.

And yet these same resentful professionals will appeal —some of them—to the same ensamples and not just as ensamples. I have often said, in courses in ethics when the question comes up as to where the philosopher can get his ethical facts and principles, that there are at least three needful sources: a) one's direct perceptions of value in concrete particular cases and in such cases abstracted and formalized, b) by deduction from metaphysical, logical, and religious premises or systematizations, c) from the perceptions and decisions and situations of the past as selected and focused in myth, poem, play, and novel. If one says these principles can be got from the *Nichomachean Ethics* and the *Treatise* and the *Fundamental Principles of the Metaphysic of*

Morals, I answer: To be sure, but only in so far as Aristotle and Hume and Kant got them from some such sources. What they add preciously to the same perception the poet may have is the recognition and elaboration of the philosopher.

The solution, so far as one is needed here, seems fairly simple. It is not the case that the original creation came to the artist as revelation, or by subconscious imagination or prediscursive insight, the artist having no doctrine or idea or access to idea or doctrine for which his plastic or analogical imagination should seek out a symbol. Nor is it probably the case that the author of the good symbolic work of art has his abstract meaning and his reader firmly in mind and then picks out his symbolic vehicle. This is done, of course, and usually the result is less than excellent. If the meaning is sufficiently simple even if not universally accepted or generally practiced, it can be the pre-existent and overarching meaning while the story proceeds, if the author is of the proper meditative cast with a still-dominant concrete imagination. I should put *The Scarlet Letter* here. A thesis which we all know (the virtue of openness and the peril of hiddenness), which is doctrinally obvious and almost unavoidable, and yet which we flout and often disbelieve in our real belief as well as in our action, is manifested and enforced by a tale at once simple and subtle and powerful in its own right and in its telling. But I think not many have succeeded where Hawthorne did; and did Hawthorne always? And I have the right to add that I do not consider the fable of *The Scarlet Letter* even undebatably symbolic or allegorical; it is rather a literal instance. Hester Prynne is very real and neither she nor Arthur Dimmesdale is symbolic though each may be called a type of moral fate; the letter A that Hester wears is a symbol.

More usually—or is it more rarely because only then it is at its best—the artist creates a character and a story and learns something of the "meaning" as he goes along, as I have said I think Hamlet grew in the creation. And what is Hamlet? I have heard so many interpretations most of them "wrong only in that they cannot help thinking themselves right." I would give an interpretation which would validate many, not all, of the others and some of those still to come. Shakespeare took the story of the Prince of Denmark and the ghost and, as he created that most real person and his fellows, the genius—not only dramatic genius—of Shakespeare constantly learned and followed and made more of the problems and solutions Hamlet makes. I think the play is the story of a powerful mind of considerable learning and great rhetoric, who is in a tough spot, and whose rhetoric plausibly presents the proper action according to the proprieties of Elizabethan England, of Aristotle (from whom he virtually quotes), of chivalry, even of Europeanized Christianity; but whose acumen, feeling, wit, and wisdom are always too wise to be really led by his rhetoric. The commentators assume he ought to have swept to his revenge, and then argue ingeniously as to what prevented him. I am not sure Shakespeare thought so and I am sure Hamlet, although his rhetoric sometimes assures him of it, never found his mind or his will or his soul clearly assenting. He had a commission without commitment, not because he lacked the power of commitment but because he was too smart and too wise not to suspect the goodness of the commission.

A symbol, it seems, cannot be only a symbol. It is not merely that there must be something for it as a sign to signify, it must have some sort of being in its own right before being used as symbol. But there is a difference in the sort of prior reality. "Being in its own right" may point to a full natural existence (as the recorders were there and playable before Hamlet used them as symbol of his voice and mystery and as Napoleon was a living man before he was ever used as a symbol of glory), or a logical being (as the triangle taken as symbol of unadjusted love and the ratio of the diameter to the circumference of a circle taken as the irrational rule of the world. "Being in its own right," however, may point to a mere "token" existence, made up simply to serve as symbol. Such are the symbols of the alphabet, of any algebra or calculus (as a little drawn triangle is the symbol for the triangle, pi is the symbol for the ratio of the diameter to the circumference of the circle, $=$ means "equal to"). That is, what is used as a symbol may have been made for that purpose, or not.

But the division may mislead us and the left-out third may be the most important for our present consideration. In art, notably in literature, everything is "made up," and then possibly used as symbol. The question is: How far was it made up in order to be used as symbol? How far was it made up to be as though having a being in its own right before any use, symbolic or otherwise? How far was it made up to serve a use—entertainment, literal instruction, edification—different from and maybe prior to the symbolic use if any? How far was it made up with a complex of motives? I believe it is the second rubric gives us the best art—and probably the best symbols.

And surely if we are to learn—as I have said I as a student of ethics want to learn—from a created story and character, we can do so, in the way I mean, only if the author's insight into reality enables him to create a story more real or more understandable than mixed-up actuality. If his story is made up by him only to symbolize a theory he already had, I shall do better, and should prefer, to get the theory from him straight—or I think I will go to Aristotle, Hume, Kant, Schopenhauer. It is as though the physicist or the biologist appeals to evidence constituted by a story he tells and from which his own hypothesis must be extricated. This is not altogether bad, by any means; the scientist does this, in part, and so does the philosopher, to explicate and test the plausibility of his theory. But factual evidence must come from the independent outside. It is this, we know, that caused the emphasis in scientific method on the devising of experiments for *future* doing. The antecedent reality of the writer's story is the correlate of the futurity of the scientist's experiment.

All communication—at least all normal communication—makes use of some vehicle, usually language either as verbal or symbolic in the narrower sense and all this is symbolic in some sense of being "semiotic." The "tokens"—written shapes or audible sounds—may be said to be signs for the words and the words for their meanings. (E.g. the two words "let" have the same tokens but stand for two meanings.) What meaning is is itself a desperate controversy, but one we can shun. We merely note that what is meant by the words may be a symbol for something beyond itself; and these are most of the symbols of art, religion, psychology. We also note that while we call the vehicles of ordinary language words and add that they are symbols of a sort, stipulated signs; the vehicles or some of the vehicles of logic and mathematics we firsthand call symbols.

That these mathematical signs are purer symbols, and yet from the literary view weaker symbols, is, I suppose, shown by the fact that, although they can like tokens be translated into words, they do not have to be, and still more by the fact that what they stand for is less "realistic" or substantial or experienceable. If we ask what are the objects meant by the ordinarily mathematical statements of modern science, we have a different set of questions in train than if we ask this of ordinary discourse or literature or religion. (Music, of course, as communication brings new puzzles.) When Hamlet calls for the recorder we know the musical instrument the word means and think of it as an actual part of our familiar world even if it is then at once used as a symbol for something further Hamlet or Shakespeare means. And this something further is not back to words again but something thought of in human, dramatic terms even though it may be, and in artistic and religious uses usually is, less concrete and familiar, cloudier and larger, more laden with metaphysical pathos, than the symbol-thing, the recorder, we started with (otherwise why run to symbolism?). Shelley's "dome of many-colored glass staining the white radiance" puts before the mind a thing and a set of qualities, must do so to be the definiens part of a simile. And the definiendum is "life"; as that of the recorder is a slight but still, to Guildenstern, an opaque "mystery." But with $f = ma$, a central statement of physics, what is the object or objects meant by it or by its terms?

It used to be supposed that we can expand the "variable" symbols, here just initials, and say "Force is equal to mass times acceleration"; and that force, mass, and acceleration, like the recorder and the dome and its colors, are known to us in ordinary experience, and that the two constants, "equal to" and "times" are known by Platonic or Cartesian reason or Lockean intuition. Well, I still think there is something in this old and naive supposition, but not many others do and even that would not really get us out. A modern physicist or logician will say f, m, a, $=$, and the adjunction of two variables are to be taken as parts of a well-worked-out formal system. If we want to say this statement, or any like it, is true, we must set up operational definitions of f, m, a, and then measure actuals. Our measuring is sometimes charged with internal and external relativities, our fundamental definitions are sometimes charged with being circular. Engineering or any "prediction" about things of the world requires the setting of "boundary" or "state" conditions and prior positions and velocities.

All this is thorny and I make no pretense of elucidation. We merely note again the unique achievement of physical science while also noting that what is meant by the symbolic statements of developed science is variously interpreted or shrugged off. And my momentary point is that there is at least a real sense in which it is science (the most developed science as proof) and not poetry or religion, which can claim to be or must confess to being symbolic.

Bishop Berkeley already pointed to the questionableness of such words as "force" and "faith" and at first wanted a symbolic reference to actual sense "ideas," later sought a symbolic reference to purpose and use. Some nineteenth and then some twentieth-century physicists discovered parts of what Berkeley taught; and Jeans and Eddington, Einstein somewhat differently, suggest that the constructions of science "are true of" but not representative of "sense data" in their appearances and orders. Others would retain some realism of a "model" and think of a world of four-dimensional, multi-geometrical space-time, subatomic particles, "wavicles," or fields of force.

In the phenomenalistic reduction, our familiarly lived-with things have no real role; and in the model reduction, our familiarly observed and enjoyed qualities have no real role. Indeed, as the things in the model world have become such things as not even our dreams are made of —very unthingy things—so the perceptual world to which the phenomenalist says science must conform has become a pallid one of pointer readings and relations— very unqualitative qualities.

Perhaps we may simplify toward some comparison and conclusion. Science as well as common sense, literature, religion, uses some sort of sign tokens. The nonformalized language of common sense, literature, religion uses tokens which are at once taken as words, words that are grammatical constants, like "and," "is," "if," or substantive constants, "Napoleon Bonaparte," or substantive or adjectival variables, which in words are almost always limited in their scope, like the common noun, "recorder." These last may then be individualized: "the recorder," says Hamlet, "this pipe," holding it in his hand. This it is, this pipe, which now is used by Hamlet, as reported by Shakespeare, for a literary symbol to bring to the minds of those who heed some further "meaning." Some of the partial variables have references more like those of science. " 'Tis as easy as lying," Hamlet says. And it may be commented that such a non-thing as lying is seldom used as a literary symbol; although of course, there may be a story about lying which serves as an ensample or allegory and so in our present usage is a symbol.

The tokens of scientific statements, we have said, can

be translated into words but need not be. At any rate they stand for constants and variables, as do those of ordinary language, but with differences. There seem to be very few substantive constants; perhaps time-space, centimeter, gram, second; and these seem queerly insubstantial. For engineering the variables can be applied and so in a sense may be said to be individuated; but there is more than a touch of illusion in this, for the concrete actual thing never appears in the scientific statement as such even when applied to an actual situation. The proper name is extrascientific; the common noun disappears early; then the adjective. It seems that as science develops, the entities that are the values of its variables, the remnants of its substantives, are more and more purely of the sort that Berkeley, at first to his distress, noted as having no idea, no thing, no ostensive definition. Not only the marks on the page, but also that for which the marks stand primarily and secondarily are symbols. The concrete, actual, existential, substantial, moving and changing and enduring stage, the intermediate and central stage of ordinary discourse which is presenting a symbol simply is not. It might be hazarded that science is as a whole a symbol for use or for the understanding of the actual; but we do not need to do so.

It is thus science that is essentially symbolic. And thus it seems to me odd that literary and religious men, some of them, should say, sometimes a bit wistfully, that science has taken over literal truth and they perforce must play up symbols. Let them rejoice that art and religion with common sense are left the possessors and guardians of literal truth. I like to believe some literal truth can be kept in science—but it is the least "scientific" part. And I willingly concede there is much symbolism in religion, more in art; but it is "adscititious and ancillary," a tool in their endeavor to win and communicate insight. When I read about the River Wye I think it is the River Wye, although I may well think of my own "wanderer through the woods." When I pray to God I pray to God, not to a father image. Better than that would be the stock of a tree that Isaiah talks of, which is at least a thing.

4

Good Use
and the
Use of "Good"*

My Topic Is the latest in the series of engagements fought over "It is me" between the purists of the *Sunpapers*' letterwriters and Professor Malone,[1] and my text is a sentence of Professor Malone's in (I believe) his concluding letter to the *Sun* of August 16, 1956: "From the earliest times to the present the rules of grammar have been based on the facts of usage (how else could they be determined?)."

I hasten to say I know the multiple letter writers and Professor Malone are able to take care of themselves and I do not want to take sides—much less do I want to make peace, which would be a deprivation to the

* To the Johns Hopkins Philological Association, February 21, 1957.
[1] Kemp Malone, now professor emeritus of English literature.

Sunpapers and their readers. And I am not presumptuous enough to come to the Johns Hopkins Philological Association and talk to you about a particular point of usage or grammar or a general theory of usage or grammar, except in an underground or adscititious way for which I have some double sort of excuse: because as a student of ethics and logic I am concerned about directive rules in general and especially about morals; and because as a part-time newspaper copyreader I am concerned about, let us say, some facts of usage that seem to me from time to time bad facts of usage and some rules of grammar that seem bad rules of grammar and some bad usage of good rules.

I have been impressed at some of the meetings of this association, as I have been impressed in conversation with students of English who have come into newspaper work, with the orthodoxy and firmness at present of the faith that a descriptive grammar and vocabulary is right, at least that prescriptive attitudes are not only wrong but also deplorable—or should I say not only deplorable but also wrong? This is a variety of the assurance of "naturalism" and "empiricism" in recent philosophy and is associated with the convention that modern science owes its triumph to its telling "how" but never "why," and with the emphases on the facts that, as Professor Boas has shown, the Mona Lisa has been admired for 450 years but for different and sometimes contrary characteristics, that it is wise to drive on the left in England, and that the Spartans taught their boys to steal.

Let me make two notes in passing. (a) If one's attitude is that prescriptive grammar is mistaken and deplorable, then one must base the deplorability on something, at least some fact of usage, beyond grammatical usage, and it may turn out to be hard to carry both the "deplorable" and the usage doctrine all the way through. (b) But there is no logical requirement that if one is a thoroughgoing usage man in grammar he must be so in everything—e.g. in morals.

Surely it would be hard for anyone acquainted with the history of morals and fashions—or acquainted with the day-by-day carrying on of any human doing that combines science, engineering, art, and practice; or acquainted with the history of science, art, language and literature, medicine, and games—not to be proportionately acquainted with the badness of—the interference and suppression by—prescriptive rules or at least the usage of rules. But it is, I take it, not needful for me to urge this side to you.

But if "prescription" is a tyrant, is anarchy possible? I am asking whether there are logical or dialectical reasons (reasons of self-evidence or contradiction) which make a thoroughgoing usage doctrine impossible. I think there are reasons that make it certainly impossible in all fields taken together (that is, philosophically, metaphysically, and I find axiologically). It is impossible pretty certainly in any large and relatively fundamental field (like language), and perhaps in any field of any sort (say in table-setting or dealer's-option poker).

It need not be, but it may be, helpful to note that a

frequent and prompt use of description is prescriptive: we are told to follow usage. And I have suggested that offenders among rules are sometimes the more obviously usage rules. There are rules lingering in the reverent rapidity of the *Sun* composing room which go back to Bill Moore—a masterful managing editor who died some time before I went to the *Sun* thirteen or fourteen years ago. Why Mr. Moore wanted it so, if there was a why, no one knows or cares. This is *Sun* style. Protagoras, the first great descriptivist and relativist, may be attacked by calling his theory impossibly fluid, or by calling his practical program impossibly wooden. Emily Post may grant it is only convention which side of the plate the knife is laid but be rigid now and unchanging in time as to putting it there.

My examples, especially the first from the *Sun*, have a double edge. No reason is given for the rule but use; but by now it is the rule that sets the use. Pure use, pure rule starting as opposites are apt to fade into one another; a world of pure use, as we try to imagine it and if we imagine it as sustaining itself, is not far from a world of pure rule. Rules codify use, but to codify is more than to repeat; and rules partly stabilize, sometimes unfortunately stabilize, use. And I think this interdependence of rule and use can go on without being either empty or blind because both rule and use, but more massively use, is not quite blind but is in some part a matter of better and worse, of choice in light of the world in which it takes place and the purposes it serves.

Or let me walk around and up on it. The bearing of the description of the prescriptive uses to which descriptions of usage are actually put will depend on what answer we give to the question: Why should we follow usage? We might, for example, accept an absolute moral or categorical imperative always to follow usage, so that after this one not-to-be-questioned rule we can be mere usage men from then on. Or we can be a sort of hedonist and assert that experience shows us that "as a rule" following usage is the best way to keep out of trouble. Or we can try to be usage men all the way through and say there is no reason we should follow usage it is simply a descriptive fact that most of the time we do.

But let me remind you of my particular text—Mr. Malone's parenthesis: "How else could [the rules of grammar] be determined [than on the facts of usage]?" "Based on the facts of usage (how else could they be?)" This, it seems to me, at once suggests another question: How is usage based or determined? And Professor Malone's rhetorical question seems to have the implication that usage is determined by nothing, except perhaps other usage; for if usage is based on any further considerations then "the rules of grammar" could be said to be based on that on which usage is based. This will be most important if the way in which usage at any time is determined by something beyond usage can be said to be right or wrong, better or worse, more or less successful. Then the rules of grammar could seek to base themselves on those considerations on which usage tries to base itself but in respect of which usage may fail of success. Even so, sagacity and experience will warn us that our continuing reference had better be to usage; but this is because use is the test of choice, not because it is its only creator.

Thus the usage of tennis players (and I have seen it change several times) tries to be that which will win matches. The rules of tennis form (i.e. the books of advice) are normally "descriptive" of the "usage" of the writer of the book or of the usage of the more fashionable and successful of contemporary practitioners. Nevertheless the writer would hardly admit that he was or ought to be just describing how tennis now is played; he thinks he has a basis in those conditions in which the successful player has a basis—in relevant facts of physics, ballistics, meteorology, instrumentation, anatomy, and psychology plus the stipulated materials and aims of the game of tennis.

Thus we ask is usage in grammar, construction, and vocabulary, pronunciation, and inflection based on anything, and is the basing capable of or subject to any order of preference? Now I am a great believer in chance (as most scientists and scholars and sophisticates are not). I believe chance is in much of our talking. Nevertheless, I cannot suppose the way we talk is altogether and purely chance. If there is determination of use what is it?

I believe in natural causation, so-called natural law whether it works from past to present to future or whether it is the formal patterning of what has happened after it has happened; and I believe that much of what we say and how we say it is caused just as is the fall of an apple or the digestion of our food when that is undisturbed by our follies or our curiosities. But not all, certainly. And even those anxious followers-afar of science who hanker after a complete causal determinism have to accommodate their faith to the facts of life so that in that curious realm of language there is apparent purpose and intended relevance, and this is enough for our argument.

Some of our language is chosen, now or through habit from an earlier choosing, and, if chosen, then chosen for a reason, express or supposed or associated; and, if chosen for a reason, then, I should say, for a reason more or less good, and certainly chosen more or less successfully with regard to that reason.

Now this is rapid and may seem dogmatic. Maybe I can appeal, in Aristotle's "dialectic" manner, to the descriptivists themselves, who quite normally, when confronted with some apparent outcomes of straight usage, say (as Professor Malone said in the course of last summer's exchange), "Of course, we mean good usage" or "the usage of educated and careful speakers." But what is "good usage?" And "educated" in what and "careful" of what? Of usage?

One might image a sheerly chance or causal first usage and thereafter no licit change except perhaps unchosen additions for new experience and pure causal shifts in phonetics. But descriptivists are normally patrons of change in language, and here I am with them. But I think some change is better and some worse and indeed many changes are actually bad and some actually good. But this good or bad cannot be in the change just as change, which is equally in all, or in the usage as usage, which also is in all. Something else is at work.

I also agree with the descriptivists, that the attempt to label or control changes in language ahead of time or in full course is apt to be all wrong, is apt to make a fool of the legislator, and is very unapt to succeed. I do not agree it is based on nothing but delusion or that because it is heroic it is silly.

I have put these considerations in terms of language, but I think they are quite general. Science is at once the most certain and the least cognate. I think no one has ever asserted that the history of science is altogether a history of fashion, that the criteria of the present scientist are nothing but the usage of his colleagues. But it has often been said that science should be nothing but a description of fact, of "how" things are, without any further explanation or assertion of "why." Science has sometimes even tried to do this, but not often and, I think, never successfully. The second scientist in our history, Anaximander, had the judgment to see that if any sort of explanation is wanted it must be in some terms beyond those of the appearance being explained. As our older "model" of explanation develops faults, we complain of it as an *ens praeter necessitatem* and discard it, meanwhile building another. The purest theory, with the least "model," must be the most formal and thereby in a way the furthest from the concrete actuality before us. And the great question of the theory of science is as to the nature of the tie (there must be some and more or less single) between the more acceptable formal theory and the reality in the fact or in the mind.

In art we now seem (but have not always so seemed) to be freest to be sophistic. "Beauty is in the eye of the beholder" is taken to be not only true in the ear of the believer but really true. But if the artist is impatient of prescription he is apt to be even more so of usage. He will probably deny, and probably rightly, the supposition that his reasons can be discerned, stated, arranged; but that he does choose, not groundlessly, and more or less well he is very unapt to deny. Some choice, I think, notably in art schools and cliques, is freely arbitrary. But even the arbitrary must survive some possible adverse considerations. Must we not say that some pictures, some music, some poems, some scenery, like some soups and steaks, are chosen, for reasons, better or worse reasons, and chosen more or less successfully? What we need in order to save ourselves from pedantry is not to try to deny this sort of rational objectivity but to remind ourselves that it is only in artificial and imposed things like judging dogs in dog shows on points that we can approach authoritativeness in our criteria.

What is the situation in ethics—the basic, old-daddy field of ethics? Well, ethics was certainly in bad shape as a school subject in this century up to about twenty years ago but since then has been a very lively one; and before that, at least since the publication of G. E. Moore's *Principia Ethica* in 1903, it was showing in the upper stories originality and vigor. And it seems to me we have now reached a point of some accomplishment in this half-century of debate in that two theses (my colleagues would not all agree) have been established for all the warring parties. (a) A pure subjectivism, and almost as surely a pure usage doctrine, will not do. It is inconsistent either or both with itself and facts of human moral experience which are too repeated and massive for anyone to ignore or deny. And (b) the objectivity of the adjective "good" defies real definition in other terms, either of non-value description or of other value words.

These two considerations have sent a good many into Moore's doctrine of a "non-natural" quality of goodness, indefinable but intuitively recognizable and present in all good things. But it seems hard for even the adopters of this doctrine to be comfortable in it and impossible for some others. I think it is close, but misses. And so I have come to say, and believe, that "good" is an adjective with objective denotation (it applies or does not apply to actual items either as they are in themselves or as they are in actual situations including persons) but connoting no single quality, natural or non-natural. It may be said to mean an imperative relation lying from the descriptive qualities of the object, and its situation, and the character of the chooser to that person's choice. But these are obscure and dangerous words; and as a sort of definition, although they help us I think, they falsify the adjectival nature of the word "good."

It may be, and I believe is, that some, perhaps many, principles or characters of a high degree of simplicity and universality (not just generality) can be found which as such are good or bad, so that things or actions of which they can be asserted can themselves be called good or bad—but only in this respect, not as-such. Thus, I think, kindliness is always and as such good, and lying always and as such bad; but the same concrete action may be kindly lying. These "universals," however, do not define "good" or "bad" or "better" or "worse" or "right" or "wrong," and no finite (or probably infinite) collection of them can exhaust the variety of good and bad things and actions. There will always turn up goods which are so regardless of what we think or feel about them and yet which are good not because they possess a common character possessed by all good things and not because they lie under any few or many principles of goodness although this they may do.

Now this defect of a single or focusable quality of goodness, I have come to see (or believe I see) is not merely a curious theory forced by threatened difficulties. It corresponds with a matter of positive practical importance in morals and one seen by certain religious moralists, notably Paul, Luther, Edwards, Wesley, who are insufficiently heeded by philosophical ethicists. These men discerned a certain special danger of badness in the endeavor for goodness. He who does good things because they are good is headed for hell. "Sinners of the left and the right," Martin Luther says (or is supposed by Richard Niebuhr to say with something like this meaning). These are the natural wrongdoers and the anxious do-gooders, and it is easier to pluck the natural wrongdoer than the anxious do-gooder from the burning.

Now all these men—Paul, the Pharisee at the feet of Gamaliel and the persecutor of the Christian heretics; Luther, the zealous and overzealous monk; Edwards, the intent Puritan youth; Wesley, the "Holy Club" member at Oxford—had come into the grace of God not from natural sinning but from scrupulous correctness, correctness which had satisfied them the less as they had pushed it the more rigorously, until their conversion, their turnaround, their, as they saw it, being turned around.

In my language, the natural wrongdoer cheats because he wants the money or enjoys winning, gets drunk because he likes to forget his cares, whores because his body is vigorous or his imagination is lustful, murders because he gets angry. The anxious do-gooder plays by the rules, stays sober and celibate or monogamous, and aids those in need not because he wills those things in their own right but because he has taught himself to believe he sees behind their actual characters some further character which he calls good and has resolved to observe. All this is not only of no virtue it is of vice. It may be this careful one is doing good so as to get to heaven or decorate his own soul (Edwards) and this is other-world hedonism and selfishness. But this is not necessarily so and still the set of the character is bad. The most resolute Stoic, with his "Only goodness is good," looking for no next-world reward, is still violative of the truth, because there is no such real character as goodness and he should remind himself that "one thing that is never good is goodness," for the essential rule is that one must do what is good because it is what it is and not because it is good. It is good, or can be called good, only because it ought to be done for its own sake; it is not the case that it is to be done because it is good.

Can we transfer these results to good usage in language? Or I should say can I transfer them, for I suspect most of you are at least not ready to accept them fully for ethics. I don't think we can transfer them just so—and I don't think we have to or want to. But a saving grace of the for-its-own-sake outcome in ethics does apply to choice in language for it applies in every field in which an ought or a good appears. There is always a difference between objectivity, integrity, honesty and hypocrisy or prostitution. As we move away from pure ethics we can become more familiar and simpler in our general answer, because the added complication and subordination of the situation allows us a simpler answer—in the same way as a private under the eye of the sergeant has a simpler choice of action than Robinson Crusoe on his island. For every more specialized sort of activity does have some known or arguable or real-though-unknown end or ends, and here integrity can in a more familiar way mean observance of those proper ends. This is why we can speak of art for art's sake less dangerously than of morality for morality's sake—though still with a double danger, for we would do better to think in terms of for the sake of whatever we think the proper ends of art are—we would do still better to think in terms of the particular picture we are drawing or poem we are writing.

Painting, tennis, music, science, shoemaking, presumably even so various and natural and profound a thing as language, should be directed each to its proper end. Of course, for sufficient reason, any item of such a field may be used for a foreign end, its external utilization being without necessary prejudice to its goodness, or badness, in its own right. So a Mozart concerto may be used to ease a mind disturbed without changing the goodness of the music; but, if a piece of music is written with the deliberate aim of therapy rather than music, the musical purist is apt to be wary of its musical goodness. Tennis may be played to fight depression, but if the depression has destroyed the feeling that playing tennis is fun the playing will be work and no play and no cure.

Within music and tennis and shoemaking we can specify some ends, and we can find, at times, superior ends outside. For ethics, by definition, no outside and superior ends can ever be; and yet no highest proper end—end inside ethics— has appeared in the long search, no *summum bonum* or even *summa bona* as a complete or completable list. Even "happiness," the catchall word, will not do. Mill "expects it will hardly be disputed" "not only that people desire happiness but that they never desire anything else."[2] I am persuaded that in literalness this is so far from true that one should rather say no one ever desires happiness, except in a sophisticated verbalism or in the sense of a more-than-generalized notion of the achievement of whatever it is one does desire. Some greatly desire to be rid of unhappiness, but that is different.

So it is that in ethics—in human morals and reflection on them—the precious doctrine of for-its-own-sake has misled great philosophers, at least has misled the interpretation of their teaching: as in the case of Aristotle and happiness, the Stoics' "Nothing but goodness is good," Kant's conviction that no action is virtuous that is not done because it is duty, even Paul's "Whether ye eat or whether ye drink ye do it for the glory of God," which is George Herbert's "what I do in anything to do it as for Thee." Paul's context in the tenth chapter of First Corinthians—if you accept an invitation to dinner eat whatever is put before you, have no scruples of conscience of your own but be careful of those of your host or others—is an argument for consideration of others and for freedom for oneself from external overruling of intrinsic goodness: "Eat any food that is sold in the market." And doubtless God is sufficiently more-than-generalized as

[2] J. S. Mill: *Utilitarianism*, Ch. iv.

that the goodness of which is inclusive of any good, God is sufficiently unspecified and beyond happiness (a mental state) or duty (a satisfied law or obligation), and yet at the same time sufficiently allows the specified or particular goods of the actual things aimed at (not sucking out from them some supposed goodness-in-itself as does the Stoic maxim). God is thus sufficiently sufficient for those of us who like him to let us repeat Paul and Herbert with edification and without distemper. Yet surely the notion of doing all things, not for their proper sakes, but for God—even with those who have kept themselves to fair doings—has over and over again led to unpleasant saints and exasperated sinners. As for Kant's famous dictum I should say: only that act is right and good which is done for the sake of what it is and not because it is duty. (I do not mean to do duty in: duty is in itself a good thing, although chiefly good instrumentally; and like every good thing should be done, when it is to be done, for its own sake.) Should virtue be *con amore* then? Surely. But we are in a tough fix if we get this far: it ceases to be virtue if we do it because we like to do it; and it ceases to be virtue if we do it because it is virtue. Similarly neither horn of the old dilemma is true: what is good is not good because we choose it; neither do we choose it because it is good.

Going along with Kant's duty doctrine is another which I think does a good deal to rescue it, and which will bring us back to language: the doctrine that there is nothing good in itself but the good will. I used not to like this either; I still do not like the words and atmosphere. I used to say a will is not good, it is right; the objects of the will are good. But now I find it coming into my own view in this way. We cannot find a common quality of good; but we can find a common quality for good wills: that they choose right. This does not enable us to go backward to a definition of good, because the choice by the good will does not make things good; but it does give us a sort of operational definition, a possible answer to the man who insists on our telling him what sort of things he ought to choose: choose what the good man chooses. I don't think the good will is the "only thing in the world or even out of it" which is good in itself—our lives daily are full of things good in themselves—but the good will is the only thing which is at once a thing and with a definite characteristic attributable to it in virtue of which it is good in itself. At least the person, or the soul, is such. I do not like the word "will," in part because it leaves out the component of knowledge, vision, acquaintance. Put this in and talk in terms of the man rather than "will," and I find a doctrine of Aristotle, which, as in the case of Kant, is associated with the doctrine I found fault with, the more widely accepted doctrine of happiness.

I used to wish Aristotle would be more definite, conclusive, culminative in his ethical writings. The *Nichomachean Ethics* suggests you better go on and wait for the upshot when you read the *Politics*. The *Politics* suggests you have already had the more fundamental answers in the *Ethics*. This is not the last word or perfect, but I now think it is part of the unsurpassed sagacity of Aristotle. And if you want to know what should be the rule of our pursuit of the good, "at which all things aim," let me read you, not from the ethics but from the *Rhetoric* (1364b 12-23):

> Again, that which would be judged, or has been judged, a good thing, or a better thing than something else, by all or most people of understanding, or by the majority of men, or by the ablest, must be so; either without qualification, or in so far as they use their understanding to form their judgment. This is indeed a general principle, applicable to all other judgments also; not only the goodness of things, but their essence, magnitude, and general nature are in fact just what knowledge and understanding will declare them to be. Here the principle is applied to judgments of goodness, since one definition of 'good' was 'what beings that acquire understanding will choose in any given case': from which it clearly follows that that thing is *better* which understanding declares to be so. . . . And that is a greater good which would be chosen by a better man, either absolutely, or in virtue of his being better: for instance, to suffer wrong rather than to do wrong, for that would be the choice of the juster man.

At the end this has a clear echo of Socrates in the *Gorgias*, and yet the general thesis seems to be the anti-

Socratic one of an appeal to numbers. Aristotle is more apt than Socrates to think that what "always or for the most part" is believed or chosen is acceptable as being rightly so; but he does not say this makes it true or good; and here there is the constant proviso of understanding and in the end of betterness. And it is not a taking refuge in authority since the rule can well be put in the first person: That is better which I will really choose supposing I have "acquired understanding" and am better. I am repeating my own thesis that understanding will always choose what it chooses because that is objectively good in its own particular right; but that, if we insist on looking for some single quality of goodness which all these things "have," the best we can do is to say each is what understanding, the good man, would choose; and that this is not just empty, any more than Aristotle's backing and filling between ethics and politics is, because we come much closer to having a sort of focused feeling for what the good and understanding man is than for all the particularities of good things. (The passage quoted from Aristotle is in the midst of perhaps the longest and most hospitable of lists of good and better things.)

Aristotle says the principle is quite general. So to a point it is, but we can get some benefit by noting where that point is. In physics as in ethics we can seldom (only for exceptional reasons) do better than take what the most understanding men say, although in all fields we have to use our own understanding at least as to why we suppose so-and-so to be the most understanding men. In each case the understanding chooser chooses according to his vision of the objective eligibility of what he chooses. But in physics the criteria of choice (even if mistaken) are isolable, generalizable, and completable; in morals (if I am right) they are none of these. Thus it is that Aristotle adds "what would be chosen by a better man" (something of Kant's "will" in the Greek intellectualism) which he does not add in the case of physics.

All this seems to me important for the transfer to the morals of language. For language seems a sort of intermediate between the limited fields of determinate interest and the illimitable field of ethics. At least in its naturalness and manifoldness, language requires its student to be closer to the student of just-any-action-for-any-end (which is ethics) than is required of the student of any other context I can think of. Even art is comparatively an artificial and narrow interest. Thus the good of the good choices of language can be like the good of the choices of action anyway—objective but translatable into no common quality which makes them good and from which authoritative rules can be deduced. It does not follow that choices are not good or bad, or that we cannot be mistaken: the more indictably mistaken when we choose the bad, perhaps the more importantly right when we choose the good. So the notion of "good usage" or "good use" is not as question-begging as it sounds and Aristotle's "good and understanding man" is called for.

And this may serve to qualify my sermon to make it

seem less smug in its prescriptive assertion. Yet the study of the uses of language is not ethics; language is not just any action for some end or other; it is in its proper nature, even if only in part, for certain definite ends; it has roles to play from which desiderata can be and are argued for and observed, roles in which still more certainly failings and failures can be criticized. What are some of the more apparent and familiar roles?

I suppose that because I am a student I naturally think language is primarily a tool of knowing. Beyond rhetoric is truth. This is what made that great rhetorician Plato suspicious of rhetoric and indeed hostile, even when he is carefully trying to do it justice, which is not always. Language, as at least a part-time worker for knowledge, has logic as something of an obligation. And this is the place to read you a passage from my son's (eighth-grade) English textbook: its passage on "It is me." I think it does credit to contemporary schoolbooks and makes some answer to the ordinary blaming of wooden rules on schoolmarms and schoolbooks.

Like any living language, English does grow and change as it has grown and changed for centuries. The changes occur slowly. There may be only half a dozen in one person's lifetime. Moreover, the changes arise out of habit and custom, not out of logic. For many generations there has been a strong psychological urge among English-speaking people to say *It is me*. This expression is at variance with the established pattern of using a nominative after forms of *be*. Yet it has been so persistent and has been so widely used by educated persons speaking and writing carefully, that it has finally become accepted.

The question which naturally arises next is: What about *It is her*, *It is him*, and other such expressions? Logically, if *It is me* is acceptable, these other forms are also acceptable. However, since language does not change logically, these other forms must win approval one by one: they have not yet done so.

The rules of grammar and usage which you find in textbooks are intended as a description of how language is actually used. The usage comes first; the description of it follows. The rules are not more important than the language. Actually no living language can be confined within arbitrary [or non-arbitrary?] rules.[3]

I like this; but in part it reminded me of one of my newspaper sentences: "They could not let political considerations override military necessities"—which made me comment that a reporter of another persuasion might have written: "They should not have let military considerations override political necessities." (Here, too, political and military are impinging systems with their own ends, raising questions of relative command or of

[3] John E. Warriner: *Handbook of English*, Book II (New York, 1951), pp. 71f.

outside adjudication. Just what is meant by "The rules are not more important than the language?" And these political and military systems, too, largely run and change by usage, propriety, not "logic"; yet their professors are much less descriptivist and anti-prescriptivist than the professors of language.) I like the textbook passage; but as in most things I like, there are points I can use in complaint. Here are two of them.

The first is the repeated idea given in the statement "Language does not change logically." This is one of those statements safe in some meanings and false in some. Nothing changes logically, even logic; for change is always of substance and logic is not substance. And if reference is to change in which consciousness and intention play any part, then I must say I have grown always more impressed with how constantly and intricately infected with logic human action is even when it is supposed by the actor to be directly reactive or sensational. And if much of this logic is, of course, bad logic; well then there is some hope of remedy in logical criticism, better logic, or logical renunciation of logic. Our passage falls into a pet contemporary propriety: *It is me* has been a "strong psychological urge" and so does not "abide our question." Surely some psychological urges have their logical or illogical, and hence criticizable, grounds. And it seems very illogical to deny or bar even the conscious use of logic in action. We do not digest by logic but if we believe we should not eat starch we can refuse potatoes. If we are persuaded it is unseemly

to split the parts of a verb (I am not) we may shun split
infinitives. These are indeed old-fashioned syllogisms.
But these sorts of logical control do not here concern
me. My question is as to prizing the logical capability
of our linguistic tool: do we, can we, should we sometimes use or refuse in language because the usage thereby
set up or retained is judged to be of advantage to the
language as a logical vehicle: that is, in the expression,
elucidation, and calculation of meanings, in the answering of all those questions that Miss Hatcher[4] has taught
us all statements are meant to answer.

One familiar party-at-interest here is precision, the
ability of the language to make distinctions. It is equally
familiar that we are always making words and always
blurring them so that we often seem to have too many
words and a fuzzy language. Here we are again in your
backyard. But I would interpose that even as a logician
I see that what is sometimes called a logically perfect
language of one word for each meaning and no more
would be unlivable, useless as a logician's "object language." Still it often seems too bad always to be killing
off the distinctions we have cut out for ourselves. And I
may add a couple of trivial—and perhaps therefore the
more significant—examples from the newspapers, which
assuredly are among the top culprits here. Especially
my own craft, the headline concocters, wreck words'
edges, for they have to make something not altogether
uninformative within the irrelevant rule of a given

[4] Anna G. Hatcher, now professor of Romance languages.

length. If the word is not too far off in construction, somewhere close in meaning, and the right number of letters, it is the word for us. From the heads the rounded words work into the body type. I deliberately used the phrase "top culprits" just now, for "top" is now one of the top culprits. That is, we not only round off all the words of a group of related meanings, we take fast hold of one (the short one if the influence is coming from the headline writers) and forget the rest. Then it is used not only in place of any of the others but where none of them is called for. I think it will be hard nowadays to find in any newspaper an account of any aggregation of government officials, business executives, military officers, labor leaders, which is not of "top" officials, executives, officers, or leaders. If the origin is in the reporting, the one-for-all word is apt to be of at least two syllables. For some time now "entire" has done the entire job for "entire," "whole," "all of," "complete." The dress shop advertises "Entire stock for sale," which should warn the shopper she need not go in unless she is ready to make a bid for everything in the shop in one purchase. When I first noticed the usage, I tried to fix the differences and use "entire" only when called for. Then I gave up and now my pencil takes out all "entires" entirely, sometimes giving a replacement, sometimes none. These pressures at times create words, especially on the sports page which is under a looser verbal rein than we are but also even quicker to stereotype. Regularly for a while and still generally anyone in sports who signs any paper, especially a contract, "inks" it;

but he never does so in any other news, much as we headwriters may want a word which is one precious unit less than "sign."

When I started reading copy, "that" was recessive; "which" having taken over as relative pronoun, and the conjunction "that" just suffering the copy reader's pencil in the few cases a reporter might heedlessly use it. Now "that" is back—mostly where it does not belong. If we want to say: "President Eisenhower said that in the beginning of the Civil War both sides lacked trained troops," we can say it so; or we can say: "In the beginning of the Civil War, President Eisenhower said, both sides lacked trained troops." If we say: "In the beginning of the Civil War President Eisenhower said that both sides lacked trained troops," we are saying something else and something temporally impossible. But more and more this last way is the way reporters are saying it. So, too, we are more and more getting quotations which keep the "that" of indirect discourse with the quotation marks and the first person but not the second of direct discourse. The two examples given here are from actual copy and, as is the case with all that follow, are not from try-outs or cub reporters but from established and sometimes eminent members of the trade or from wire copy. (I avoid adding quotation marks to these two examples of quotation marks.)

Stewart said he told the President that "I will do my level best to live up to his trust."
He said that "we are watching with a great deal

of interest" what other censorship states were doing with their respective statutes, since the Supreme Court recently overthrew bans in Ohio and New York.

There is evidently a "strong psychological urge" to write this way. Does that settle it that it will be, must be, ought to be "accepted?"

I said I had two caveats to the passage from my son's English text. The one was against the denial of logic, the other is against an ascription of logic. "Logically, if *It is me* is acceptable, these other forms [*It is her, It is him*] are also acceptable." Logically? Well, this will depend on whether there are any differences within the personal pronouns which are relevant to the case used after the verb to be. Whether there are or not is no question for logic but it may be for metaphysics. My onetime and longtime English teacher, Professor James Wilson Bright, was a more unqualified champion of "It is me" than is Professor Malone and he used to say such a use of the pronoun is "an absolute use" and that "every other language than modern English—schoolroom English—has had sense enough to use an oblique case for the absolute use." Metaphysically, is not the first person singular more absolute, perhaps the only one that can really be? However that may be, I do think there is a different feeling with which we use the words that refer to ourselves—I should say with which each of us uses the words that refer to himself, especially the personal pronoun. If someone asks "Who is there?" there is a radical difference, probably of semantic reference and grammatical meaning, certainly of assurance with which I can say "Me" from the meaning and assurance with which I can say "He" or "Him" or "She" or "Her." Indeed, I can express my meaning with a "meaningless" noise, satisfactorily to myself and also to the inquirer if he knows my voice. I must rise to a more semiotic level if I want to communicate the meaning of any of the other pronouns, even the first person plural.

Metaphysicians, then, may be said sometimes to have a right to intervene in the too-hasty logic of our rule makers and revisers—if not in the common sense uses of natural grammar (since common sense is the basic metaphysician, which is why metaphysicians can learn and have learned so much from grammar and grammars). If so, metaphysicians might try—but how?—to protect some of their own words. I know it is futile, but I go about complaining of the fate of "category." This might have come under my first rubric of blurring and stereotyping, for it is hard now to read or indeed hear any reference to sort, class, kind, group, set, species, genus, family, phylum, variety, tribe, type, cast, denomination, stamp, description, brand, make, stripe, strain, style, color, range, grade, heading, caption, title, rubric expressed by any other word than "category." Last week I read of a category of tugboats used for some purposes in the inner harbor. Yet this is one of the reverend, the high and mighty words, parsimonious of application, one of the most needful if also difficult words of metaphysics and philosophical logic for 2500 years. Within a few weeks it has become a tired little maid-of-all-work.

But language is much more than an instrument of knowledge. It is, for example, a sort of dress. Are there not good and bad ways of dressing, no matter how much scope we give to personal taste? Pretentiousness is an easy habit—certainly in newspaper writing. It has something to do with "entire." Many reporters, especially if they have any business connection, will never say "more than"; it is "in excess of"; "before" is "prior to," and "about" is "approximately." There is some affectation of nonpretentiousness in the present insistence on the contractions for the verb to be and all auxiliaries. These could be but normally are not ambiguous; but it is hard to think of any reason for them, at least outside reported conversation—and not much there. Very few words or phrases are pronounced as written: why not imitate speech more widely? But then we need a new spelling for each city—for Bawlmr and Trontuh and N'Orlns along with the standard "she'ds" and "I'ms" and "they'lls" and "it'ses." Then the wire services would be sunk, and I understand the contraction fad comes down to the writers by directive from the executives as part of the unremitting effort to enliven the style. From above also, I believe, comes the loss of the article from the beginning of a sentence. "A," "an," and "the" are short words, so the "logic" seems to run, and hence are not strong; although both the logician and the linguist are apt to think them, the syncategorematics generally, especially potent.

Language is not only a mnemonic for oneself it is a preservator of the past beyond ourselves. We can welcome change and still want some continuity of principle and some retention of the ability of our ears to hear the literature of other generations. Must we lose all prepositional constructions in long pile-ups of qualifying nouns? These certainly are often hard to unscramble but I find their offensiveness more direct than obscurity. Here is a headline—not, I am glad to say, from the *Sun;* but from our flighty little sister, the *Evening Sun*—a three-column head, which ought to give room for a little more resemblance to English: "P.W. Talk Rule Change Rumors Cited By Korea Flyer in Germ War Probe." If it be said headlines are a lawlessness unto themselves, here is a phrase from supposedly important wire copy: "taking part in National Institutes of Health mass influenza vaccination studies"—and the wire copy comes without the help of any capitals. The offense is compounded when we pile-up as titles all the adjectives and appositional nouns that go with a person's name. Danish Nobel Prize Winning Physicist Niels Bohr appears on the same page with Movie Actress Grace Kelly and Jewish Lawgiver Moses.

Clarity, precision, adequacy, decorum, historic integrity—something like these and others like these, I want to say, can be believed to be goods which usage should seek and against which usage may offend. But always there is the more that makes these goods not absolute, that adds other goods not specifiable, that puts language near ethics, and swings us toward ethical anti-nomianism and objectivity. And this returns us to our sinners of the right and left.

Our general problem of use and the good has, it seems to me, got itself a new situation in this century. We used to be able (before we were so bent to do so) safely to be descriptivists bowing to use and welcomers of change because there had always been two chief sources of change and both good. These were the wise and the ignorant; those who knew and loved language, and those who cared not at all except for its direct use; scholars, writers, historians, and ploughmen, thieves, children, longshoremen, race track people. Sometimes the one became a schoolmarm and a sinner of the right; sometimes the other was vulgarized into a sinner of the left. But generally both sorts did the "living language" good and there were no other real innovators. But the more essental sinner of the right has now greatly multi-plied and has more power over our uses of language than his predecessors: those who care much and profession-ally for the utilization (rather than use) of language but for extrinsic ends and who neither know nor love language itself. They are of the right because they care-fully follow and make rules for the sake of a reward— which, indeed, since the reward is extrinsic, they may well get while the poor moral sinner of the right is going to hell. I mean of course many (surely not all) of the publicists: newspaper men, advertising writers, radio and television personnel—many of the writers and talkers, more of those in authority. I do not want to be unfair; certainly not to my own craft. Most of the newspaper men and women I have known have been devoted to the honest getting and telling of the news; and this is their primary mystery. Many of them have wanted to tell it well, and some have made of this a true and by no means unstudied devotion. But the diversionary, and sophisticating, and depressing forces playing upon them are strong—like those playing on the lawyer in Plato's *Theaetetus.* So it may be that "having no soundness in him," the worst of it is that he "is now, as he thinks, a master in wisdom." Certainly he is sometimes now, especially on television, the master of our language, the one to whom people turn to imitate or imitate without turning.

I tag on a miscellaneous collection from copy. If I were not one of the poorest of note takers I should have many times as many to select from.

Certainly when the Charles Center and the Civic Center came up as projects, everybody's ire, including the press, rose and everybody had something to say on it.

. . . reached its peak when an admitted jewelry store holdup man was pardoned.

The A.A.P.A. is the ideal agency to spearhead any drive to iron out these difficulties and would build up its own prestige (which is needed) at the same time.

A C-124, the military designation for the Douglas Globemaster, yesterday landed at Friendship.

With the new oil concessions, the role of Lake Maracaibo undoubtedly will rise to become an even more vital segment in the world's offshore drilling picture.

Holiday deaths took ten lives.

A man with twenty years of tubeless know-how.

There has been difficulty from time to time conforming the types of personnel needed by the Port Authority, a completely different type of agency than the usual under the State jurisdiction, but most of them were working out.

Too much dependency upon one person is never too wise.

Highlight of the cardinal creation ceremonies will take place Thursday.

I hope I need not add that I am sure the reader of this paper will have found sentences he would willingly put in this concluding list.

5

Proprieties and the Motion of the Earth[*]

WE LECTURE ON the coming in of the Copernican astronomy and we are familiar with the supposed—some obvious—reasons people resisted the notion that the earth moves. We fail to be impressed by the twentieth-century resistance, less conscious and more successful, to any notion that the earth does not move. And we do not notice that the reason for our putting aside any notion that the earth stands still is the same reason which made the sixteenth century abhor the notion that it moves.

We are familiar with the logical motives of the Copernican argument: the principle of simplicity and the principle of relativity. It could be maintained, beyond this, that the principle of relativity, properly descriptive and not normative, when it grew clearer as to its depth and

[*] To the Johns Hopkins History of Ideas Club, November 6, 1952.

scope in this century, came to add a prohibitory aspect; at times seconding, at times overbearing the more properly normative principle of simplicity, and serving as the front-man for another but unexplicit principle, that of nonuniqueness, essentially normative, but lacking the factual and logical force of the principle of relativity and falling short of the simplicity of the principle of simplicity. That is, relativity (the principle) tells us we cannot expect to measure certain supposed motions but does not give us any basis for choice among theories except among theories of what we can and cannot measure; simplicity tells us of two rival hypotheses to prefer the simpler. Both seem logically and historically safe. Nonuniqueness, more complexly and psychologically founded, tells us to shun any hypothesis which gives man a unique place.

When the Michelson-Morley experiment measured the velocity of the earth as zero,[1] scientists did not bother to argue against the chance the earth's velocity is as measured, absent. They went to work to explain the result.

[1] Or close enough to zero to be accepted as such, plus "experimental error," by all except Dayton C. Miller. For Miller see his "The Ether Drift Experiment and the Determination of the Absolute Motion of the Earth," *Review of Modern Physics*, v (1933), 203-242. For the experiment see W. M. Hicks, "On the Michelson-Morley Experiment," *Philosophical Magazine*, III (6th Series) (1902) 9-42, 256, 555f.; and in A. A. Michelson: *Light Waves and Their Uses* (Chicago, 1903). The role of the existence of the ether in the interpretation of the outcome of the experiment will be noted later.

It will be said there are empirical proofs of the earth's motion, if not measures of it. Even some scientists sometimes say this. A recent, evidently authoritative, and to an outsider most interesting book says of the pendulum experiment: "In 1851 Foucault devised an experiment that established once and for all that the earth rotates on its axis." And of the orbital motion: "The final and ineluctable answer to these questions was provided by Bradley, the then Astronomer Royal, in 1727" by means of the aberration of starlight.[2]

But in other contexts it is a commonplace with scientists that this finality of proof is not so—not so of motionlessness and equivalently not so of motion. They know that in the best information and logic of physics it cannot be so. I think the same fallibility of evidence could have been known in the time of Copernicus.

When in 1543 the *De Revolutionibus Orbium Coelestium* appeared, it carried a preface, since subordinately famous. Unsigned, it was soon discovered to be the work not of Copernicus but of Osiander, Lutheran theologian, who saw the book through the press. And for the preface its "well-meaning" author, from that day to this, has suffered the reproaches of all (so far as I know) who have commented.

The anonymity of a preface so placed is obviously blamable; but this chicanery is only incidentally the cause of the reproach. The theme of the preface is this:

[2] Sidgwick, J. B., F.R.A.S.: *The Heavens Above;* American edition prepared by Warren K. Green (New York, 1950), pp. 35, 37.

If, however, they wished to weigh the matter thoroughly, the author of this work has done nothing for which he merits censure. For it is the job of the astronomer to use the careful observation of an artist in gathering together the history of the celestial movements; and then—since he cannot reach the true causes by any line of reasoning—to think up or construct whatever causes or hypotheses he please for these movements, provided that by taking them as postulates he can correctly calculate from the principles of geometry the past and future movements of the heavens. This artist is especially outstanding in both of these respects. . . .

For it is sufficiently clear that this art is absolutely and profoundly ignorant of the causes of apparently irregular movement. And if it constructs and thinks up causes—and it has certainly thought up a good many—nevertheless it does not think them up in order to persuade anyone of their truth but only in order to institute a calculus correctly. . . .

Perhaps the philosopher demands likeness for the truth instead; but neither will grasp anything in the way of certainty or hand it on, unless it has been divinely revealed to him.

Therefore let us permit these new hypotheses to make a public appearance among old ones which are themselves no more like the truth; . . . and, as far as hypotheses go, let no one expect anything in the way of certainty from astronomy, since astronomy can offer us nothing certain; for if anyone were to take as true that which was constructed for another use, he would depart from this discipline more stultified than when he came to it. Farewell.[3]

Now, the relativity that Osiander appeals to is the relativity that the Renaissance learned from Aristotle by way of Simplicius, the relativity that comes to prominence in the troubles ninety years later of Galileo with Maffeo Barberini become Urban VIII (did Galileo teach the motion of the earth as something more than the mathematical hypothesis the Pope had prescribed?). This relativity, as a mathematical but not physical or metaphysical doctrine, was used by Galileo's generation vigorously, as earlier it had been used lightly, as an exculpative for scientific theory. But in Osiander it seems (or is this just to the Einsteinianly prejudiced ear?) to be used somewhat more in the modern and honest spirit. And to be sure it had been used quite honestly, we now know, by astronomers before the year of Copernicus's book and Osiander's preface. This is that astronomical relativity or provisionalism which the age called "hypothesis." The associated, but still historically and scientifically distinct, principle of relativity—the difficulty or impossibility of assigning motion to each of two bodies which are observed to be moving in respect to each other—the principle which had been

[3] Reprinted from "Great Books of the Western World" by permission of Encyclopedia Britannica, Inc.

noticed by Nicholas of Cusa and which was to be formulated more precisely by Galileo, this too was used by Galileo and Kepler and the other Copernicans quite honestly and strongly in the negative against the arguments of the "Aristotelians" and common sense. It is a curiosity, indeed, how Galileo and the others could use the principle of observational relativity honestly in one direction to explain away the empirical evidence urged by their opponents, and use the relativity of "saving the appearances" and "hypothesis" to guard themselves from doctrinal reproach, and match these two uses with two feelings toward what were associated principles while using neither one against themselves.[4]

And Francis Bacon, who is often decried as enviously cool toward Copernicus, asked something better than Copernicanism expressly because he was one of those "philosophers who demand likeness for the truth," because he felt neither the Copernican nor the Ptolemaic theory could give more than a relativistic "saving of appearances."

[4] This is in part, perhaps, why the detailed and negative use of the principle of relativity continued, and continued as scientifically respectable, while the more general form of it as a view of the nature of astronomy, if not of all science, was first satirically used, then denied and forgotten almost, before being revived by Einsteinian physics. Here, too, are proprieties. For a summary and clarification of the at-times controversial doctrine of "hypothesis" in the Renaissance (which was not available when this paper was written) see R. M. Blake's Chapter II in Blake, E. H. Madden, and C. J. Ducasse, *Theories of Scientific Method* (University of Washington Press, 1960).

There is no room to expect any pure truth from these or the like theories: for as the celestial appearances are solved both upon the suppositions of Ptolemy and Copernicus; so common experience, and the obvious face of things, may be applied to many different theories; whilst a much stricter procedure is required in the right discovery of truth. For as Aristotle accurately remarks that children, when they first begin to speak, call every woman mother; but afterwards learn to distinguish their own; so a childish experience calls every philosophy its mother, but when grown up, will easily distinguish its true one. In the meantime it is proper to read the disagreeing philosophies as so many different glosses of nature.

Before this, but in the same fourth chapter of the Third Book of the *De Augmentis Scientiarum*, he says:

The pains have been chiefly bestowed in mathematical observations and demonstrations; which indeed may show how to account for all these things ingeniously, but not how they actually are in nature: how to represent the apparent motions of the heavenly bodies, and machines of them made according to particular fancies; but not the real causes and truth of things. And therefore astronomy, as it now stands, loses its dignity by being reckoned among the mathematical arts, for it ought in justice to make the most noble part of physics. And whoever despises the separation between terrestrial and celestial things, and well understands the more general appetites and passions of matter, which are powerful in both, may receive a clear intimation of what happens above from that which happens below. . . .[5] We, therefore report this physical part of astronomy as wanting, in comparison of which the present animated astronomy is but as the stuffed ox of Prometheus—aping the form but wanting the substance.

And in the sixth chapter of the "Description of the Intellectual Globe," in the course of his longest (and a shrewd and interesting) discussion of astronomy, he speaks of Copernicanism as "savoring of the character of a man who thinks nothing of inventing any figment at the expense of nature, provided the bowls of haphazard roll well." One thinks of Einstein's "God does not play at dice."

Osiander claimed relativity for the Copernican theory, supposedly as an exculpative from popular or clerical

[5] It is usual to say Newton unified the worlds above and below the moon by extending the laws of falling bodies to the planets. This sentence and others ("For that supposed discrepancy between the celestial and the sublunary bodies appears to us a figment at once driveling and presumptuous"; "Description of the Intellectual Globe," ch. v) are then a clear expression by Bacon of the ideal that Newton achieved. The different feeling in Bacon, however, suggests we might more accurately say Newton extended the celestial to the earthly, made the laws of falling bodies an instance of heavenly regularities, or that the laws of falling bodies are properly astronomical laws. Bacon would make the heavens earthy, perhaps.

blame; Bacon charged the Copernican theory with relativity and called for something "nobler" and more "physical." Both are reproached.

What Osiander is accused of is not so much anonymity as it is cowardice—especially making Copernicus seem cowardly. That the theory was simply "hypothetical" and "calculative" was denounced as false of the intention of Copernicus and of his followers; it was taken as false of Osiander's own belief and to have been intended by him to avert persecution by the fast-hardening intolerance both Catholic and Protestant, of which Osiander felt himself more aware than the dying Copernicus—somewhat anxious though Copernicus's thoughts of publication had been made by the shadows of authoritarian controversy.[6]

The contemning of Osiander is probably historically not altogether unjust. But the anonymity was not as bad then as now, and the view expressed was not made up *ad hoc* and might have been Copernicus'. And did the

[6] It is significant for histories of intolerance that Copernicus, having waited, he tells us, four times Horace's nine years, decided on publication as the storms gathered. Even so, in part thanks to Osiander perhaps, it was not until 1616 that the Congregation of the Index ordered: "Lest opinions of this sort creep in to the destruction of Catholic truth, the book of Nicholas Copernicus and others . . . are suspended until they be corrected." (see Preserved Smith, *The Age of the Reformation* [New York, 1920], p. 622) Sir William Dampier says: "It was understood [in 1616] that the new theory might be taught as a mere physical hypothesis." (Sir William Dampier, *A History of Science* [New York, 1938] p. 124) Notice how Sir William reverses Bacon's meaning for "physical."

preface not say what is now scientifically respectable truth of the new theory? Is it not more than time we recognized it said truth? Might it not have been said that Osiander had some realization of the truth of what he said, not merely that it was true in particular but that it must be so; and further that Copernicus might have realized this; that he had at his command reasons for understanding it; that, if Osiander's imposition had not been uncovered, the prestige of Copernicus might have led us to ascribe to him not the timidity of avoidance but the perspicacity of genius?

And Bacon—does that "bright, mean" spirit not deserve our notice that his view of Copernicus is something subtler and more of insight than ignorance or envy, and that it may have been insight into the nature of mathematics, not ignorance, as is often said, that led him to give it only the "great auxiliary" (but not "substantial") place he several times and emphatically but briefly gave it? Would he now say to Descartes: "You won; where is your winning? Cartesian physics is slipping out of all the Cartesian imaginableness which based your physics and linked it into metaphysical reality. Quantum mechanics runs off into an arithmetic of probability beyond geometry, thrusts dynamics into kinematics and so, perhaps, demands my qualitative motions. 'Pictorial meaning' has become a jest with the positivists. If physics does not come to me, it has done with you."

And when physics is refined to a sheer mathematical relativity, what shall we do? Shall we try for something more "physical" with Bacon, or with Osiander for something "divinely inspired?" Shall we with Plato seek a "dialectic" to "destroy the hypotheses?" Or shall we say, "We can know no more"; or, "There is no more?"

Both Osiander and Bacon are reproached for seeing, from different angles, relativity in the Copernican astronomy. Yet what would the Copernicans have done in their defense against the enemies of their doctrine without the appeal to the principle of relativity?

The ways of supposedly detecting the motion or motionlessness of the earth have been: by direct observation ("We see it does not move"); then by noting effects as due to uniform motion (used against the Copernicans in the assertion that such effects—a wind past the earth, a fall of bodies to the west—should appear on a moving earth and do not appear); then by the effects of acceleration (how bodies fall or liquids behave in a jerking carrier came to be appealed to as evidence for a moving earth, but only in the guise of the next test); then by distinguishing the effects of "artificial force"—centrifugal—due to "real motion" (the flattening of the poles of the earth, the trade winds, the pendulum experiments); and then by measurement of one peculiar physical process, light, which by variations in apparent speed or direction past us may be supposed to disclose to us our own motion (the aberration of starlight, which seemed to succeed, and the Michelson-Morley experiment, which seemed to fail).

The first two were used against the doctrine of a moving earth, and relativity was the Copernicans' answer. The last two have been used for the doctrine of a moving earth, and here the principle of relativity has been called in only to explain why the proofs of motion cannot be used as measures of where and how much. The proofs maintain our experimentalist respectability. The real basis of our faith is elsewhere.

We can see that there is motion; we may be able to describe it in a system of bodies; we may be able, with two or many more than two bodies, to say that at most one of them is motionless; but we cannot distribute the rest and motion except by taking some point as fixed.

So the Copernicans explained to the man in the street who saw the sun moving in the sky. So, accepting the sun as center, they charted the paths of the planets about it. So, taking the "fixed stars" as the frame for that system, we can speak of the sun as moving. Then we see the Milky Way as moving among the galaxies, and then all moving in space or just moving. And so our success should warn us not to be too single. These beautifully charted revolutions about the sun are, as relative motion, sufficiently certain; but they are also, if one please, a going of the sun about the earth.

It is possible to construct a model, a mechanical toy, of the planets and the sun and to operate it as that system would be projected upon the surface of a table. We will first do so holding the sun in the center of the table. Then, illustrating the motion of the whole system, we

146

move the sun along the table with its planets still revolving about it. And if this is possible, as it clearly is, it must be possible so to move the sun that any chosen member of the system, say the earth, shall remain stationary on its original spot on the table. Or we hold the earth still and let the rest of our mechanical toy operate, as it will, of itself.

Even for the imagination to authenticate, this is not too difficult, especially if the fixedness of the earth be just of its central point, i.e. if the axial rotation be not stopped. If this be stopped the motions become much more complex—but we know we can hold our toy either with a pin through the poles of the earth or with our hand around the equator and the mechanism will keep on swinging without any systematic discomposure.

The difference in the motions around the sun and around the axis is important for the earth-motion arguments since, when the rest of the starry universe is taken into view, the axial motion is much the harder to still with imaginative acceptability. The vast enlargement made in the universe by Copernicus (to account for the fact he could detect no stellar parallax), one of the chief sticking points to early wonderers, becomes the affront to imagination if imagination is asked to revert to a nonturning earth; for then the whole circle of the heavens must be traveled every twenty-four hours by the farthest stars, and the velocities (not to speak of the momentums) called for, and those still greater ones called for by now-accepted distances, become stupendous. Thus is suggested a third or middle view, as of Brahe and (sometimes) Bacon, in which the earth turns but does not travel. This too would seem to be secure against disproof by observation.

The principle of relativity permitted the Copernicans to use the principle of simplicity in support of the heliocentric theory. Further nicety of observation (notably the phases of Venus) also supported the theory as against the Ptolemaic; since even if the sun goes around the earth the other planets still go around the sun as mathematically more central to their orbits. But as to the motion or motionlessness or centrality in the universe, as to this no observation of the motions which take place among our company of observable bodies can give any evidence. Of this the principle of relativity is the statement.

Nor can the principle of simplicity, in any descriptive sense within the system, have voice here. All relative motions remain as they are discovered to be. Our toy heavenly bodies will continue to move as they "naturally" do regardless of what we do with the whole toy so long as we do not interfere with its internal economy. And they can be described in any way in which they can be described, under the same rule. Our heavenly motions are assuredly better expressed in Copernican than Ptolemaic schematism. But as to the common sense opposition between the Copernican and Ptolemaic points of view—is our earth home staying in one place and is that place at or near the center of things?—no answer can come from observation, and no longer can it come from mathematical simplicity, which speaks decisively for the Copernican-Keplerian-Newtonian description of the relative motions but which, since all of this remains even if the earth or any other thing be taken as the one fixed member of its company, can speak no further.[7]

If simplicity is to speak here it must return, and does, to some other, to some more naive, dramatic, picturesque, human simplicity. Is it simpler to have the bigger body fixed? Then which is the biggest? Is it simpler to consider ourselves central? Is it, on the contrary, simpler to exorcize "anthropocentric prejudice" by not allowing the human habitat to be set apart? Is it simpler to say that any of the vast number can equally claim preference and so it is vastly improbable to assent to any one claim, especially if that claim correspond with any prior interest? Or is it simpler to deny that there is any meaning in the question as to which is unmoving?

The last is the now sophisticated position, with the next to last leading on to it. But I think it would not be contended that as yet it is the "felt" answer except among a minority even of careful scientists. I think there

[7] Kepler is the real nonrelativist. For all his enthusiasm for and his contributions to Copernicanism, he argues vigorously, on the basis of both observation and theory, to a unique place for the earth: it, and so the sun which is the center of the earth's little system, is in a sort of "hollow round" within the encirclement of the stars. This has been pointed out, since this paper was written, by Alexandre Koyré, *From the Closed World to the Infinite Universe* (Baltimore, 1957), pp. 64ff, 81ff.

is ample evidence, in addition to plausibilty, to show that it is based upon or accompanied by a conviction that it is meaningful and for all practical purposes truthful to say and to require it to be said that the earth does move; and that this conviction in turn is almost solely based upon the anti-anthropocentric pressure of modern science, which is the reverse of the pressure of the sixteenth century in face of the same consideration of alternatives.

The modern pressure can be, and should be, put with more breadth as a principle of nonuniqueness: Let no thing claim preference. Relevant to the motion of the earth: Whatever, from time to time and for descriptive purposes, be accepted as fixed, let the choice be accidental to "reality," at least let it be independent of extrinsic interest.

Notice that it is not proper to say, as is sometimes said and often implied, that the principle of relativity forbids or prevents this or that state of affairs in nature or the possible truth of any belief. It is a statement of a characteristic which our minds observe as essential to observation. As such it may "forbid" as illicit certain sorts of conclusions from observation; but it cannot prevent the illicit conclusion from being true, nor can it forbid any state of affairs which might falsify or bypass the principle and enable us with unobserved validity to reach that conclusion. If we decide to call some body— sun, stars, ether—fixed, partly perhaps because we do not see any chance of detecting motion on their part, we must not object if someone takes motion with respect to that body as real; and if we find or seek a bar to our detecting motion with respect to such a body—as we did with the ether—we ought not to feel that difficulty as flowing from the principle of relativity since the principle could save itself as easily by querying the "absoluteness" of the body.

The principle of simplicity, asserting a basis for choice, may, if accepted, be said to forbid or command among material assertions; but not the principle of relativity. On my way to the art gallery I stop and consult an esthetician and an oculist. If the first tells me that blue pictures are better pictures and I believe him, I will feel some compulsion to see the blue pictures as better. If the second tells me I am color blind and cannot see which are blue and then I do see the blue pictures that others see, I will quit believing the oculist. And I would have disbelieved him from the first if he had assured me that as a necessary consequence of his diagnosis all my friends and chance acquaintances would unfailingly contrive to mislead me as to what they saw as blue simply to prevent my finding out. Thus if an observed motion be by hypothesis an absolute motion then such it is. We know it to be such only from the hypothesis and not from observation of it, if the principle of relativity be true. But it is foolish to tell us that if we accept the hypothesis we must also believe nature adds to that particular motion some extra and accurate ingredient of falsification lest the principle be violated.

Against the sort of objection to the theory of a moving earth frequent in Copernicus's own time, that if the earth moves then things dropped from a ship's mast would be left behind to the west, that the air would be a wind counter to the motion, and that we would find it harder to walk or throw a ball one way than another, came the formulation of the principle by Galileo: In a closed system at rest or in uniform motion everything will happen irrespective of the rest or motion of the system. Acceleration is excluded from the statement. Acceleration was for long supposed to be a telltale sign, telltale of real motion precisely because the force was artificial. As Henry More somewhat contemptuously pointed out: he might be unable, just by looking at their positions, to tell whether he or his friend had run away and returned; but, if he saw his friend red in the face and panting from his exertion, he would feel safe in saying his friend was the runner and not his own cool self.

So the physicists themselves, followed by all the physically tutored world, came to explain that the side-climbing water in the twirling bucket proves the bucket is turning; and so our books can still say, as we saw, that the pendulum experiments and the flattened poles of the earth and the trade winds prove that the earth is really turning on its axis.

Yet, it is curious that not until recently was the needlessness of these conclusions noted. The water climbs up the sides of the bucket without regard to the earth's motions and changing motions in which it takes part but with regard to the rotation as between the bucket and the gravitational surface of the earth. To be sure we know ordinarily that we have applied the motion that produces the turning, but the evidence is only that the climbing of the water follows the relative turning not that it follows only the turning of one of the bodies or only the use of a particular force.

Even after the Newtonian unification of forces under gravity it was not recognized that such a description of forces operating regularly among bodies in space made the description of motions resulting from those forces as much a matter of relativity as the visual observation of respective position.

We can say to Henry More: To be sure, your friend did the running but maybe he ran to stay still. If you suppose the whole solid earth with you on it to rush away from under your friend's feet and then back again while he by hard running just manages to maintain his position in space, you need not be surprised that he is red and you cool even though you are the one who has indulged in a quick round trip. If some demiurge were to hold suspended in space a bucket of water and below it the globe of the earth, then, keeping the bucket still, were rapidly to twirl the earth right beneath it, the water, we may presume, would climb up the bucket's sides. The trade winds and the flattened poles of the earth—to most people the most persuasive "proofs"—are evidence of relatively rotating masses, not necessarily of absolute motion. And the deviation in the swing of the pendulum is no more than these.

But only today, with the Einsteinian principle of equivalence, under the incitement of newer theories which doubtless have their own shortcomings of self-understanding, are we seeing the defects of insight in earlier achievements. And here we are apt to think that the correction is to be made only if we are taking over the whole of the Einsteinian theory, whereas the correction is valid independently.

The principle of simplicity again is a major factor in the Newtonian dynamics for the Copernican astronomy; much of the prestige of the Newtonian law being in its explanation of many domains in terms of its own general relations. All bodies are falling bodies. In the Newtonian scheme, given inertia and gravity and some initial distribution of masses in motion, the whole adjusts itself as a system. The scheme does not, presumably, rule out the possibility of accidental or resultant uniqueness: within the system any body might through a balance of forces turn out to be central or stationary or both (supposing a meaning for the words). But it must be any one, not an externally preferred one. Already in 1700 the scientific mind, trained in simplicity and relativity and becoming more and more careful of nonuniqueness, is "conditioned" to suspect the suggestion of a particular body as such a stationary one; and soon will refuse to consider even the possibility of, or any offered evidence for the actuality of, such a status for this body, our body, the earth.

If we toss a handful of the twenty-six letters into a groove, they may fall into the order of our alphabet just as well as into any other one order. But if we come upon them in that order we feel pretty sure they have been arranged. The odds are no longer one against any other one in the multitude, but one, or a few, against all the rest of the multitude. And in the vastly greater multitude the universe, even if this piece of matter might, like any other one, be unique in station, this piece of matter is, we know, our home and place from which we look out on all the rest; and we refuse to believe that the adjustment in collocation of unnumbered particles would so have come out to answer our antecedent interest. If not physical law and not chance, then nothing; and our principle of nonuniqueness becomes, as against the proper principle of relativity and more than the principle of simplicity, a maxim and a prohibition in the choice of theory.

Thus the same consideration, that the earth is our home, the earth is valuationally unique for us, moved both those who fought against and, in part, those who fought for the new astronomy. And that consideration—of the concomitance of the valuationally unique and the physically unique—which leads us to reject, to ignore, the supposition of a central earth is the consideration which led the Middle Ages to accept, to refuse to question, that same supposition as imperatively proper and natural.

The realization that the principle of relativity is safe from observation of force effects as it is safe from observation of change of position should have come, we have seen, hard after the use of the principle restricted to uniform motion; on the basis that all motions may be described as functions of the presence of other bodies. Actually this realization came, and then not generally, only after the suggestion of still another modification of theory: the assertion that motion relative to the speed of light cannot be detected.

This is sometimes put as an extension of the principle of relativity: not only to deny the possibility of observing real motion but also to deny the possibility of observing relative motion of anything with reference to anything correctly assumed or reasoned to have real rest or motion. To the outsider this would seem not only to be an exceedingly queer rule but also to open the way for a run around the principle of relativity: all we have to do is find that thing with reference to which we cannot measure our varying motion and we know, by observation, what thing has absolute rest or motion, even if we cannot know—and with light we somewhat unexplainedly do seem to be allowed to know—what that motion is. Thus the principle seems to turn out to be the barrier not to the knowledge of absolute motion so much as to our knowing our motion.

For some time before Michelson first made his inteferometer in Berlin the day had been looked forward to when the changing velocity of the earth relative to that of light should be measured. To be sure the logician could point out the principle would be logically serene. We would know the motion of the earth relative to another material motion, light. If we assume light is in its own right absolute, we do so certainly not by any observation of its absoluteness; and, doing so, we already assert a measured absolute motion. Any further assertion of other motions derived from that motion which we assume to be absolute depends upon that assumption and is not any more an offensive bit of absolute knowledge than was the original. Even if we allow ourselves to say we know by this means the motion of the earth relative to the aggregation of bodies making up our "universe," is that universe moving and how?

Nevertheless, it is hard not to sympathize with the physicist in his impatience with the last question and in his feeling as to the prior caveat that on the basis of a tremendously authenticated general physics light may be taken as moving freely through the space among all the stars. So for all his purposes, if the measurement in view were obtained, the puzzle of the motion of the earth could be taken as solved, let the logician debate as he please about the principle of relativity.

Yet, out of this comes one of the basic paradoxes or confusions in the expression of recent theories. We are told that the principle of relativity is *saved* by the discovery that it is impossible to detect any motion relative to the motion of light; that light always passes any body, regardless of the motion of that body, at the same speed. And this is to say that it is impossible to detect with respect to one material motion precisely that sort of motion, relative, which the principle first told us was the only sort that can be detected.

The history of it seems that the principle of relativity

was used primarily to defend the thesis that the earth moves; which became the thesis that the earth moves but that its motion cannot be measured; and the principle came to be identified with the thesis. So it comes to include the assertion of discovery: that a particular observable body or process, light, has a unique mode of observed behavior. We must then find for this thing some unique nature to bear its unique behavior, or, better, find that this thing has a unique place in our methods of observation and measurement which will enable us to account for its curiousness as an object of observation and measurement.

This is one case in which the principle of nonuniqueness is flouted. It is so, perhaps, for the sake of stricter adherence elsewhere.

For, of course, in the Michelson-Morley experiment the measurement in view was not obtained, at least the figure obtained was near zero. If the experiment had succeeded as expected, the measurement for the motion of the earth would doubtless have been accepted and soon it would have been emphasized that the principle of relativity was not radically offended.

How could the zero result of the experiment, so often varied and repeated, be accounted for? The suggestions were many. One was not made even to be cast aside— one which might seem the most natural: that the earth does not move. We experiment to determine the velocity of the earth; our result is zero; why not say the velocity of the earth is zero? To be sure the flattening of the poles and the trade winds, the pendulum experiments, the aberration of starlight were supposed to show the earth's rotation. We were assured and are still assured we know the earth some time during the day has a minimum motion through space of more than forty miles a second. But as soon as theorists got around to it under the new compulsion to criticize the bases of the Michelson-Morley experiment, the bases of those other proofs of motion were removed.

Without intention, I have not made reference to the ether. It seems clear that the existence or nonexistence of the ether is irrelevant to the considerations in question. But, perhaps, this irrelevance should be noted, since the Michelson-Morley experiment came about in the midst of belief in the ether, since what was done and what was looked for are likely to be recounted in terms of the ether, and especially since an instance of the power of proprieties is that it became the fashion for expositors to say, when the suggestion of zero velocity was at length raised by outsiders, that the experiment was really concerned not with the motion of the earth but with the existence of the ether.

The two things involved in the attempted measurement were the earth and light, whatever either be or however either move or not move. The mathematics of the differential effects in view is the same with or without the ether. The usual (Michelson's and Eddington's) description in terms of the hindering effect on the "rower" or "swimmer" by the "current of the stream" makes us think of the ether streaming past the earth, but the calculation is identical if instead of "opposing and assisting current" we say "retreating and approaching reflectors," which is indeed the assumed situation. If we complain that we cannot be content without modeling some medium for a wave or some material traveler, we can replace the ether with an Einsteinian dynamic space-time or bring on the photon retaining enough wave-character not to behave like a baseball.

Michelson says, "It was considered that, if this experiment gave a positive result, it would determine the velocity, not merely of the earth in its orbit, but of the earth through the ether." He goes on to make clear the importance of this distinction not by any interest in the ether but by reference to the presumed motion of the sun: that is, the experiment would tell the earth's motion not merely in the solar system but along with the sun through space. "It was hoped that with this experiment we could measure the velocity of the whole solar system through space. Since the result of the experiment was negative, this problem is still demanding a solution."[8]

The experiment assumed that light is independent of the motion of its source, but this independence was one of the facts causing the belief in the ether, not a deduction from that belief, and it survives the rejection of the ether. Indeed, the ether was by some (by Michelson himself at first) called in after the experiment to take back what it had been expected to give: first, to be carried along by the earth and so impart earth's velocity

[8] A. A. Michelson, *Light Waves and Their Uses*, p. 159.

to light's; and then to help give some how and why for the Lorenz-FitzGerald contraction. Dayton Miller, who continued the experiment and the belief in the ether for many years, held that with the assistance of the ether and some carry-along he could both account for the virtual zero of result and get sufficient basis for some deduction of an actual velocity. It is the more strange that the orthodox party, who reject both Miller's theories and the ether, should almost offhand take it that by rejecting that which to Miller seemed the ex-onerator of the experiment they would exonerate it.

It is sometimes said that the word "ether" came to be used by the successors of Maxwell, for whom a material rigid and elastic medium for Fresnel waves was no longer needful, just to stand for a frame of reference to allow the passage (in the equations of electromagnetism) from one system of co-ordinates to another, i.e. to serve as a sort of Newtonian absolute space and time.[9] In this sense the "ether" is an assertion of a real space-time as against a relativistic one, and the issue can be stated so that to reject the ether is to beg the question for Einsteinian relativity. Then the Michelson-Morley experiment may be said to be set up to decide between Newton and Einstein (if these be the only alternatives) and so to decide against the ether. But this is not what "ether" has normally meant or what it means now to most hearers or, I think, most users.

The Michelson-Morley experiment, especially as

[9] Cf. Louis de Broglie, *The Revolution in Physics*, tr. by R. W. Niemeyer, (New York, 1953), pp. 8of.

leading on to Einstein, helped to depose the ether by illustrating its supererogatory character for a positivistic or formalistic or phenomenological physics, but the experiment does not seem in its own right to have been directed or to have demonstrated either for or against the belief in the ether, nor does the elimination of the ether affect whatever implications the experiment may have for the motion of the earth.

Perhaps the only thing that would have persuaded scientists to weigh the possibility of taking from the earth the motions ascribed to it would have been the repeated performance of a Michelson-Morley experiment on some other body with positive outcome.[10] In the absence of that, it is assumed that the experimental zero got here would be got also on the sun or the moon, that the performance of the experiment here is the equivalent of the performance of it anywhere.

This last represents the broad, deep feeling we have called the principle of nonuniqueness; the wariness of any preferential standing in nature, the refusal above all (a sort of uniqueness in nonuniqueness) to make the earth unique among its hosts of companions, to make this of all the bodies in space motionless—this the one on which we observers chance to find ourselves, the one which the older theology told us must as man's home have a central place in God's creation as in God's interest.

In the theory of relativity we have an interpretation

[10] Or on an artificial satellite (1960 note).

of the experiment in terms of the nature of space-time or of measurement. It may be we have the right to say that all observed motion is relative not only to other body but also to some accepted or master means of measurement, and observation. Thus all statement of motion requires correction for the messenger, a factor which cannot be corrected for but must always be a factor.

Suppose Baltimore and London were synchronizing watches. Telegraphic messages are used with correction for the time taken in transit. But is the time in both directions the same? Another correction should be made: for any increase or decrease in the figured time of transit due to the destination's coming to meet the message or running away from it. But this cannot be told. If we set up our Michelson-Morley apparatus in Baltimore and London we find the messages go by at the same speed. And since we cannot measure how and how much each is wrong in setting its watch we say each is right; or so we are told. Herein it is taken as true (a) that light in any direction passes all points on the moving earth with the same speed, and, seemingly, (b) that there have been different times required to traverse equal distances in different directions across the surface of the moving earth.

How is this last known? Even assuming that light has a constant speed in its own right (and some suspicion seems to be cast by the theory on our means of getting at this absolute), some of the paradoxicality which people find in the theory is in the joint assertion that this constant speed shows up as unaffected by the speed of passing points, but that it is to be taken as not unaffected by the speed of passing distances: all points go by light at the same speed but distances between points do not do so.

If, as an alternative supposition, the fact that light rays go by Baltimore and London at equal speeds be taken naively to show that they took the same time to make the trip in either direction, then no indeterminable correction would be called for. Light would really act across a moving earth as if the earth were still. (This would be equivalent to the theory of a dragged-along ether but without the ether.) And the theory might be more consistently loyal to the motive of relativity. But without the dragged-along ether, this theory would seem useless or incongruous in the Michelson-Morley experiment and with or without the dragged-along ether it would seem preposterous for some ways of stating Einstein's famous embankment story of the two rays of light coming from opposite directions to the midpoint of a moving train and to the midpoint of an embankment which were opposite one another when the two rays left sources at the two ends of the train.

It is from the embankment story rather than from the Michelson-Morley experiment, I suspect, that the majority of us outsiders are apt to get most of our impression of this "restricted" portion of the relativity theory. And the outsider, it may be, can find some incongruity or puzzlement between the impressions he gets from the two. From the embankment story he gets the notion of different elapsed times which make a difference and an observable difference—but the different elapsed times cannot be corrected, because of the identical passing speeds. In the Michelson-Morley experiment there is no observable difference. Although the experiment was designed to measure passing speeds it does so (as all measures presumably in some way must) by means of distances and elapsed times. If the elapsed times of the two perpendicular journeys were not equal or somehow brought into equality, the result of the experiment would not have been zero. But on the embankment it seems clear that the point of observation, on the train or on the embankment, which is closer to one source of light when it is observed will then be some distance from the point on the other with respect to which it is moving, and there must be some interval between one observation and the other. It may also be supposed that one observer can observe the other and with the two rays from opposite directions can observe the contrast of one simultaneity and one before-and-after or of two before-and-afters with reversed order. And here no literal FitzGerald-Lorenz contraction will remove the differences, which are all along one direction. (Not that the differences will tell us which if either is standing still.)

It is the single ray in either direction in the embankment example that requires the two-timing of simultaneities. And we can refuse both the singleness and the rays and have something more like messengers walking along train and embankment; and then there will be no two-timing. We can ask for two rays of light in each direction, one for train and one for embankment: could we do anything with them? And the two-timing would seem even surer to return with distant sources of light. How can we justify light going across distance with differing speeds when the two lights are not yet aware of the differing speeds of the bodies to which they are going or (somewhat less preposterous) instantly adjusting their speeds to the speeds of the neighboring bodies they happen for awhile to traverse? Imagine two stars, one east one west, signaling in such a fashion as to be called simultaneous on the east-west embankment. Can they be so on the train?

This suggests that on the train and the embankment we could ideally use two Michelson-Morley experiments and could actually do so if we had a fast enough train. Even a one-armed apparatus would be enough, within its accuracy, to show the relative motion. So Doppler effects familiarly show that, although light may pass two points at the same speed, pulsations, and thus periodic signals, it will not pass relatively moving points at the same intervals.

We had better stop teasing the concrete illustrations or stories offered for our help by the physicist. But we can say to the already impatient physicist that he does offer them, that it is by them he attracts much of the general interest he does attract and at least partly likes, and that when his physics calls for revision of common sense and metaphysical beliefs it must in part be tested by them. The arguments of Henry More and Newton are not more absurd and disposable than they are cogent and available.

Whatever be the proper expression for its effect on the classical principle of relativity, the theory of relativity is said to ascribe celestial motions to the distortion of space-time by the presence of bodies. "Geometry" being the systematic analysis of the order of a manifold, and a "space" being any manifold of which a geometry is possible; then our universe may be described as in or of a space with varying geometries, and the behavior of things can be attributed to the change of geometry from region to region; and the behavior can be explained systematically if there be discoverable, as there must be in a world of "natural law," any consistent basis for the alteration, as in the presence of mass. So light bends as it passes the bulk of the sun.

Here, just as the principle of equivalence of gravitational and centrifugal gives the statement of the unproving of the proofs of the motion of the earth by means of the pendulum, the bending of light gives the statement of the unproving of the second of the proofs quoted at the beginning of this paper. The aberration of starlight seems a return to the observation of change of position, but an observation of direction relative to that unique thing light, as the Michelson-Morley experiment tried to reach speed relative to light. But the bending of light, even beyond that calculated from the corpuscular theory, early served the theory of relativity not only as something to be explained but also as something to give evidence for the theory. And, indeed, that the aberration of starlight could be and would be explained on relativistic grounds could be taken as clear, ahead of the particular explanation offered.

It has been noted that the sophisticated assertion is not that the earth does not move or that it moves, but that "moves" so simply used is out of place. And it may be felt this vitiates much of what has been said. But much of our concern has been historical and in this connection the objection is irrelevant. And today it is still the case that scientists, if not most of us, do say the earth moves and say this in a way they do not say the earth does not move; and they, if not most of us, do resent any intimation that the earth stands still in a way they do not at all resent straight assertion that the earth moves; and we do all this, it seems, on the same ground on which people four hundred years ago made the opposite commitments.

Someone says, "I see the sun move, not the earth." We say, "You see, from one of them, the motion of two bodies relative to one another." Someone says, "Bodies do not fall to the west as they should if the earth is moving east." We say, "They and the atmosphere are parts of the earth's system." Twice the principle of relativity works in rebuttal. Someone says, "The poles are flattened by the spin of the earth." We say, "Science gives empirical proof." We do not say, "All are parts of a system and the effects of a turning are of a turning of one thing among a company, relative as is the change of position we see." Someone says, "We measured the change in the speed of light past the earth and got a zero result." We now say, "All are parts of a relative system and anyway we have to expect a zero result in deference to the principle of relativity."

The bases of our faith are in none of these facts but are the same bases the Middle Ages had, but used inversely. They are, if you please, metaphysical (I am not one who dislikes the word). And the metaphysics of a time is important—especially the metaphysics of the physicists of the age immediately before that in which physicists are in dominant esteem. Our physicists today are relativists; but, sharing the faith that at least and assuredly the earth does not stand still, the one application of their relativity they are most apt not to have taken the trouble to make is the unproof of their predecessors' proofs of the motion of the earth.

The Greeks seem to have had little of either the medieval or the modern jealousy. Anaximander taught a free earth and the atomists a traveling one in a wandering universe. The Pythagoreans' earth moved, but close around the center. Aristotle went back to a central and still earth and Aristarchus on to a heliocentric scheme. Epicurus, elegant economist and relativist in science, said say whichever you like so long as you keep it "natural." Ptolemy reached back to Aristotle. Each of these and others chose for his own reasons and without facing a conditioned uproar of response. We seem less free.

For my part, I am enough of a modern to prefer the modern use of the principles involved to the medieval use; but not by a margin I cannot imaginatively cancel. Like most people today I am apt to feel that the earth is moving—that is, I feel so in my more but not in my most scientific moments. In my least scientific moments I take the earth as still; and in my most scientific moments I say there is no absolute motion in proper meaning for physics.

But also in my more careful and cantankerous moments I am fond of annoying people by saying, honestly, that there is not demonstrative proof or experimentally decisive evidence that the earth moves, and that what experimental evidence there is might seem to say that the earth stands still.

6

Sua Si Bona Norint[*]

"WHAT THE COUNTRY NEEDS is more thought about theology and gambling," he said. He is a theologian and poker player. "Say on," I said. He did.

The class war, that is our passion now. The phrase is arrogant; "a class war" would not be so bad. There are no real classes of humans, no one privileged classification. It is true that close to two fundamental classes are the "right men" and the "wrong men": those who have been sufficiently skillful and lucky and those who have not, winners and losers—as you can see if you watch the two groups going home from the dice game. They sometimes fight; but hardly as groups and only if they are bad gamblers or there is cheating. I suppose it is this right-men-wrong-men division that gives au-

---

[*] This paper was written in the latter thirties when fascism-naziism and communism were both flourishing, when it was possible to talk with half or full partisans of each, when some others were feeling the pinch of a supposed forced choice, and when this rancor flavored the longer-dated and more amenable division of right and left and of rich and poor in the English-speaking world.

thority to our right-left debate with its sometimes insistence that the game must end in a fight, and all of us must choose sides. But I would say that the elementariness of the right-men-wrong-men division is evidence of the elementariness of fortune or luck in our lives, and that its relevance to the right-left debate suggests that therein we ought early to decide what we mean to think about fortune and what we mean to try to do about it.

To be sure, life, in part a game, is not just a game like a dice game. You cannot take it or leave it alone. The rules and the objects and the players are largely forced on you. You have to play with the cheaters too; hence the police. The conditions change, however, and the rules, never all they should or could be, ought to change also. That usually means some sort of fight. But the bad temper is always more than is needed.

Theoretically the leftists are now in favor of abolishing the game. They think economic life should be a severe business of production for use, and all gambling is bourgeois frivolity and wickedness. For the rest of the world the Communists rely on stirring up fights within the game. This is the conscious duplicity Moscow has taught. And their converts are largely made for them by the unconscious duplicity of their opponents, who avoid seeing that the contest in which they have been winners is a game at all, who talk and act as though it were a heaven-ordained discipline and an infallible test of the ability and the virtue of men. I want to keep the game with clear recognition it is a game of luck as well as of skill. I want the left to say what they mean and stop adding injury to insult. I want the right to stop adding insult to injury and to say and mean other than they have been saying.

In the seventeenth century we fought about theology: say about supra- and infra-lapsarianism. A little more attention then to social economics would have been helpful; at least as showing that even theology is not the whole of living. Now we are in as much and a very similar froth about social economics. Here is a phrase I saw: "the liquidation of class solidarity by the bourgeois ideology." That is better than many, for it does apply "liquidation" to "solidarity," an unusual aptness; but still it will match anything from the Synod of Dort where infra- and supra-lapsarianism led the brethren to battle. The Arminians and Socinians had more feeling for language than the Marxians and their big-business comrades in combat. But the point I wanted to be making is that, along with our delusion that social economy is the only important or real interest, we have made for ourselves another delusion which consists in forgetting all about God and chance.

I am surely not prescribing religion as the "opium of the people"; though it may be no worse than the heroin of communism taken as an orthodoxy. I am not urging the "prisoners of starvation" to be good losers as a solution of our troubles. The chief sin is on the other side of the barricades; and I am not offering a solution. Let the fight go on; history may care who wins. But when both sides by the same error of fact make the probable outcome of the fight less truthful and good and make the fight itself more unpleasant, for themselves, their opponents, and the few if any still unconcerned, then it is abstractly well to view the error, even if it is interesting only to those few unconcerned.

Now it seems clear that a portion, worth considering, of the hard feeling between right and left, between those who differ as to private property and its division under capitalism, springs from the modern attitude of turning the head away from the reality or the legitimacy of fortune, luck, and chance. Back of this, to be sure, there is on the one side the fear which is one of the tolls of precarious possession, and on the other side the hunger of want in the presence of plenty; but these are increased and sharpened and poisoned by the mutual resentment which would be allayed by some mutual obeisance to the god of chance.

To say this is obviously to suggest some defense (not final) of capitalism and its fortunates; but it puts the greater blame upon the capitalists for their own and their system's ill-odor and actual danger. If the have-nots are poor losers, the haves are wretched poor winners; and it is much better and more needful to be a good winner than to be a good loser. He-who-has-not thinks his indignation is against the unequal distribution of wealth since he feels that he-who-has cannot prove his right to the larger share; but in this he-who-has-not

is wrong, or at least the burden of proof is still coldly
on him, since there is no more call upon the lucky pos-
sessor to prove his right to what he has (I do not refer
to the legality or the decency of the means by which it
came) than there is call upon the seven to justify its
appearance on dice after the seven has shown.

What really infuriates the have-not is the attitude of
superiority, of smugness, of rewarded merit, of proved
desert, which the haves assume. In this the have-not is
quite right; since the one indefeasible obligation imposed by fortune is humility, humility toward fortune
and the less fortunate. *Noblesse oblige:* not to prove any
bare prescriptive "right" to the favor of fortune, but,
theologically, to return piety for the gift, and, ethically,
to justify the possession in the good use of it.

Oh, of course, we all one way or another really
believe in luck; especially when ours is bad—that is one
respect in which the depression was not altogether un-
helpful. But we do not acknowledge it. Have you read
all the speeches of all the presidents "on the state of the
nation?" Neither have I; but you may be sure you will
not find in them any diplomatic recognition of chance
in its purity—except perhaps in some dim relation to
ill-health, which those of us who remember *Erewhon*
are apt to think not the most judicious choice for a
single sharp exception. Even the columnists, the political
and economic ones, avoid the unseemliness of luck in
their celebration of success, encouragement of industry,
advice to folly, discovery of error, suspiciousness of

plots, and denunciation of malignancy. And it is just this official and social acknowledgment that carries the weight, for it is the official and social acknowledgment of our own and each other's stature, merit—reputation, that objectified ghost—that fills us with present and expected elation and chagrin.

Naturally, I do not assert that there is no determinant but chance. Good luck by itself is seldom sufficient for accomplishment even in worldly goods; bad luck is frequently not enough to keep a good man down. Yet luck is the most of many a success, and in one way or another (for bad luck could always be sufficiently worse) luck is a part of all success. St. Augustine called it (in his sphere of interest) free grace; but we have reached the height of assurance where we will not even acknowledge plain facts under the colorless name of fortune. What we have we make our glory, and our glory we refuse to share or be grateful for. "We have worked for it," and in a sense this is perhaps true. Yet we might have worked harder and better and got less, and how many are there about us who have worked harder and better and got less? My convictions are capitalist. I am apt to be enthusiastically capitalist after being with the righteous wrath of Communists, especially the more "intellectual" and vocal ones; but the self-righteousness of the beneficiaries of capitalism catches me back to a lesser dread of its overthrow.

I am not arguing for a philosophy of resignation, contentment with whatever is as right. An acceptance of what is past, without repining as well as without approval, of "that state of life unto which it hath pleased God to call me," is a wise and salubrious and sporting though difficult attitude; such an acceptance of what is still to come is lazy, deadening and altogether repugnant to my thesis. Chance is still free, the future open; there is no foreworking personal luck; and it is the part of vigor to gather one's abilities with a prayer for fortune and have after the thing you wish. The gospel of chance should go along with the old-fashioned American recognition of the equality of men and declaration of equality of opportunity—hard things to keep bright and sharp.

Once it was the have-not democrats who asserted chance, who were the gamblers. The few who had were all for insurance, that is for hedging; and for governmental carefulness in the subjection of the subject. They had a suspicious eye for the "radical" doctrines of individualism, least government, free thought and criticism, and equal opportunity. They still have, where they can; and so in parts of Europe they have made themselves up a right which is a state tyranny different only in words and beneficiaries from that of the left, and just as collectivist and just as obliterative of the individual and his fortune. But now in the English-American countries, with the usual unexpected logic of history, the conservatives have the principles of liberalism left in their hands. Communist and Fascist have gone medieval and are all for being "organic" and authoritative. And the liberals in the liberal countries are more the hedgers, more for having the state insure everyone against chances as well as risks. But the conservatives are uneasy in handling the weapons of the old liberal offensive for their own defense; and they are still more hesitant to give recognition to so unsettling an adventurer as chance. So, for the most part they sit upon their money and make faces of injured virtue.

I am not arguing for a pure *bellum omnium contra omnes*, for no rules to the game, no help for the unfortunate or weak. The "natural rights" of life, liberty, and the pursuit of happiness, whether natural or rights or not, are at least negatively well established. It is the function of government not only to keep order but also to interpret to the social conscience of the time the rules for the conduct of the economic game, protection of the underdog, and an increasing minimum of security for him. We may look for the time when no one will be in desperate need materially, while still holding it important that there shall be opportunity to have more than one needs. Doubtless such fellow decency is also economic shrewdness. But apart from this consideration of providing the greatest possible wealth for the game to be played with, the love of the game itself may leave us both willing and anxious to eliminate desperation from its lower fringes. A good gambling house, I am fond of noting, will give you carfare back to town and a bit for your pocket.

Something might be said (ignoring that orthodox

modern silliness that the economic interest is the only real or significant one) for the proposal to eliminate chance from our economic life leaving its play elsewhere. But wealth, and money, make the naturally best chips. Let us then, as rapidly as may be, eliminate from the economic gamble that hazarding of actual life and health which has too often been forced upon too many of the losers; let us be reluctantly willing, for this purpose, to limit or tax the upper winnings; but let us cherish the game. Is it not at least a permissible ideal, to look for a good game with the state doing as little as possible beyond enforcing rules as simple as possible and assuring something to those who win nothing; as an ideal more appealing than that of a state-owned level allowance to all, or of that sweet denunciation of all gambling "from each according to his ability, to each according to his need"?

I am not attempting to argue here the economic case for capitalism or against communism; but suggesting that both sides look to their metaphysical foundations and especially brush up on their gambling philosophy. For my own part, I like disparity of financial fortune for somewhat the same reason I am glad all persons do not look alike. I believe large fortunes are good things but not good things to be sought too supremely by anyone. To the getter of great wealth I would give applause for his accomplishment and that pious regard due the favored of the gods; but I would have him feel he owed some apology and some well-evidenced explanation to the public that his wealth in its greatness was the gift of luck: until then let him be suspect as one who has devoted to material and personal possession too much of the spirit of a man.

This matter of admiration is diagnostic. In the classic world there were the Fates who obscurely overarched even the gods; there were also special deities of fortune; and all the gods who intermeddled in human affairs were particular and capricious beings. Men could look upon their fortunate (or unfortunate) brethren as somehow good (or bad) since they pleased (or displeased) some or other of the gods, but not as absolutely and totally things of merit (or demerit). And men could and did feel the saving propriety, when too fortunate, of making some sacrificial curtsy to those profound Fates who represented the more inscrutable avengers of merit and presumption.

As this from-some-points unworthy picture of the gods became abstracted and then personified again into the monotheism of Christianity, the intermediate Epicurean emphasis upon chance dropped out, and the needed accounting for the puzzles of reward was saved for the time by the Pauline and Augustinian doctrine of grace. But this in turn was worked against by the moral-theological difficulties, and the theory of exact Providence, despite variously ingenious qualifications put in by more perceiving minds, made for more and more self-satisfaction on the part of the successful. Today this is intensified in our anti-theological worldly self-sufficiency, until the man who has done that holy thing, made money, reposes upon his conscious merit, placid with himself and most annoyed with those who covet; and they who have not made money rise wrathfully in their conscious merit, which they feel has been defrauded of the just return of its work or denied that which should be the portion of all without work.

I dislike the other-worldliness of medieval religion; but it did leave the poor and unfortunate of this world a hope and a pride. We have taken flight from other-worldliness, not in a spirituality which tries to know both fact and value, but in a worldliness without piety —that is without any respect for the factual powers and events beyond our control or for the values which are not our brutal drivers but our fair and difficult objectives. Job's comfortable friends assured him he must have done evil to bring evil upon him. Job's wife argued if he had done no evil then God was evil and urged him to curse God and die. The solution offered by the whirlwind God of Scripture is scarcely clear or satisfying, but it might be interpreted as an appeal to chance and a patient scepticism. Anyway, Job got well and was richer than ever. Our reds, like Job's wife, are the more alarming; but our comfortable ones, like his friends, are the more maddening. And they add more recruits to the ranks of their enemy than does all their enemy's red propaganda.

They, these comfortable ones, write bland letters to the paper, congratulating any occasional show of moderation on the part of labor organizations, and assure them that in this great country wealth is waiting each man in nice proportion to his industry, frugality, and virtue. The laborer reads and moves to the left. The comfortable ones descant on the glory of the big-business executive, his gorgeous salary and bonus, and they point to the still more gorgeous profits of his company as justifying the reward of his services, plainly irreplaceable since he is one of the very few accomplishing such profits. They should be patting him on the back as the lucky one to be holding the plum, so properly juicy; and at the same time reminding him in the words of Emerson, "Believe me, today's most indispensable, tomorrow ten thousand men can fill your place and mine."

Why cannot they at least bethink themselves that while the great executive is, perhaps with most admirable ingenuity, judgment, and flow of effort, earning his company's millions, many a man is struggling with quite equal ingenuity, judgment, and effort for some end, imposed upon him and to him imperative, of which the reward is very little, even mere escape for the moment, altogether unattended by glory? Surely, we know that the difficulty of achievement has no necessary proportion to the importance, the fame, or the monetary size of the objects dealt with. Back of the luck often involved in success, lies the luck often involved in what jobs we find at hand. "And in the teeth o' baith to sail, /It makes an unco leeway." Our Rigid Righteous are under the rubric of Money not Morals. There is room for a new *Address to the Unco Guid;* if the machine guns of a more up-to-date dialectic do not first find their more compelling if less convincing voice.

The situation is the more astringent in America because of our official tradition, connected with our discountenancing of luck, that everyone shall "work" and be rewarded in money and position. We tend to a one-talisman valuation. The scholar, the artist, to be sure, has his somewhat grudging recognition apart from his earnings; and he has the actual satisfaction which comes from his activity—as has the stamp collector, the "society man," the race-track follower. But the many who willy-nilly are only too intent on the pursuit of money and are still failing, these have only the failure and the weariness of effort spent. When along with the belief in a single sign of value there goes the belief in its unfailing award, the pursuers add to want and disappointment the stigma of shortcoming in the one all-important test, with no allowance granted for all the untold ways in which mere chance *may* have worked against them.

Thus in the economic game is intertangled the hazarding of an undue weight of mental health and social temper. Triumph and chagrin, in degree, are proper to any game; in our economic game they are made needlessly and falsely serious. Our poor men and our rich men have troubles enough; there is no reason socially to consecrate and impose upon them those Stoic "conditions of sin," grief, fear, hilarity, and greed; let us leave the rewards of business, for all their importance, where the Stoics saw they belong, as "advantages" not as the consummate and most real goods and proof of goodness. The elimination, earlier referred to, of the lowest fringes from the economic game, requires an actual expenditure and probably some curtailment of upper profits; but this elimination requires no sacrifice except the sacrifice of falsehood. Perhaps this, the effort of what Mr. Santayana has called "cleaning the windows of the soul" is precisely the hardest of all efforts to call for.

As it is, even the remedying of actual injustices in the conditions and conduct of the economic game is not guaranty either of its life or of its decent demise. Many who lose find themselves with no refuge for their self-esteem; and they make a refuge, not illogically and not always untruly, in a bitter charge of unfairness and cheating. On the other side, they who wish to be bland and comfortable grow shrill when they find their blandness derided and their comfort threatened. So we have on both sides self-righteous hurt feelings and fear—fear of future changes, fear of present insecurity. After awhile we, in between, the majority, are told, "You must choose," and "Which side of the barricades are you on?" As for me I will go to the races.

In the United States, please heaven, we are still a distance from the barricades. But if we do not want to leave our continued blessedness to chance we should give chance more recognition now, in the interest of

what should be the respected respectableness of the unfortunate, and of the fortunate.

(I have been fond of urging some use of the old
principle of selection by lot in our politics. Aristotle
would not call us a democracy because we suppose ourselves to go about picking the best; and our experience
might tell us we would often pick better by not picking. Let us, for example, in the selection of a United
States senator, in the primaries be given a slate of twenty
chosen by lot from our party and from these pick six
nominees by vote. Then, in the election, vote for four of
the nominees from both or all the parties, and from the
elected four take one by lot to go to Washington. The
benefits of this politically—I think they would be real—
are apart from the present story; but it would have virtue in popular economics at least as apologue.)

Let us regard our rich men as not merely able but
also lucky and therefore pardonable as well as admirable; not, just because of their wealth, either as reverend
or as wicked. By all means let us increase our national
wealth and its distribution; but meanwhile let us give up
the smug impiety which ties up tightly together deserving and money reward, and which sharpens all our
natural human imbecility of peevishness.

In this opposition of peevishness some men are bad;
some women are worse. And here, also, the worse performance is by some of those on the better side. The
wife of the unlucky man has even more excuse than he
for bitterness, if less than he for sorrow. But behold

(not always nor as often as seems to the bitter eyes, but too often) the wife of the lucky one, as she displays her three attributes: her wealth, her emptiness, and her assurance of high deserving. Let the Daughters of the American Revolution be after her: she makes more Communists than all the "subversive influences" in the terrible colleges.

Let the capitalists realize their position as representatives of freedom and intelligence, of courage and willingness to gamble. The day is long gone when the "upper classes" can hope for government to guarantee continuance to them under the shield of "vested interests" (phrase once sacred, then anathema, now antiquarian). But they may get themselves the *chance* to continue and some others the chance to win. Their "radical" opponents this way and that have gone seeking absolutes and oppressions and quick beatitudes. They are left in defense of the modern faith, the belief in freedom, in the individual, in criticism, in experience, in variety, in chance. These are all things having the common essence that they are liable to error. But, also, they have the gift of present honesty and cleanness, and of infinite possibility ahead. They call for humor; but humor is not fatal to them.

As for huge inheritances, I would not object to the government's taking most of them—if they are not just to be poured down the government's drains. Let them be given away in sizable chunks to other individuals by lottery. I believe we would be both better pleased and in a more acceptable state in the eye of heaven if each of us were given, not the benefits of another government bureau and not a thin dollar, but a sweepstakes ticket which may be worth—lovely thought—a million.[1]

[1] This was written in the nineteen-thirties without any international reference. Reading it over in 1961, I think of what the peoples of the world think of us as a people. Without getting into the campaign debate over "prestige," we may suppose most people abroad think of us as very fortunate and would be happier in that fact if they did not also think we take our good fortune as the mark, if not the simple result, of our virtue.

7

Bridge

"Why, What Do you like, if you hate both disputes and whisk?" said (and "with great surprise") David Hume to Horace Walpole in Paris, 1765. In America, 1932, he would leave out the "disputes," increase the surprise, and vastly broaden the circle to whom he might address the question.

All strong things have the defects of their merits,— or owe their strength to the merits of their defects. Of all things in the world the game of bridge is one of the great triumphs of the second of these classes. A gamble which is not a "good" gamble, a test of skill which is not a "good" test of skill, it is yet a skillful gamble which, escaping the repulsions of the sterner gamble and of the sterner test of skill, has held steady place in our civilization for two hundred years. Today, in its most up-to-date form with its essential defects most ingeniously contrived to merits, it makes its professors wealthy, its amateurs zealots, and itself an amazement to the few still unconcerned.

I hasten to say that I play bridge, have played it from bridge-whist through auction to contract, have won a little money at it, have had much entertainment in it, have played it more hours than I care to think of. But I have also played chess, long and arduously and ardently; as I have poker—draw, stud, and crazy. I have played many of the hordes of minor card games: pitch, blackjack, faro, fan-tan, red dog, rum, hearts, stop, banker and broker; and solitaire from the most intricate and careful to the most idiotic. I have played dice and roulette—how long!—and I know the greater game of the races. I have not been a professional nor have I had great skill in any of these, but I love them. Now, at almost all of these others I have had far more intense enjoyment than ever I have had at bridge. And yet all these I find often giving way, gradually giving way, again after a new spurt giving way, to bridge. (I bar the races, for that entrancement is so much more than a game that it would be no fair competitor here.)

The stronger gives way to the feebler, so often; the feebler must also be the stronger. Shall we not say the stronger in respect of our weakness, the weakness of the flesh, of the spirit? I prefer to play bridge when I should prefer to prefer to play poker. Then, thanks to bridge which serves my weakness. But woe is it that it encourages and seduces my weakness, and woe is me. I do not mean I succumb to bridge with constant remorse and anguish of spirit—any more than I feel delinquent when I read stretches of newspaper, stretches in which

I have no particular interest, instead of soaring with Shelley or struggling with Einstein. But then we are not told that impertinent columns of the paper are more thrilling than *Adonais* and more cogitative than the theory of relativity, as today we are told that bridge is the one master game—and more than a game, at once a perfect relaxation, a moral exercise, and an intellectual school.

Let me admit early that much of my irritability with bridge is due on the one hand to personal "tastes" which, though arguable like all tastes, are beside the point here, and on the other hand to the "atmosphere" of the game which is more or less accidental to the game itself. The major games have their own traditions, as distinct as those of baseball and tennis. In the noble game of crap (not of the gambling house but of the "friendly" variety) I have sometimes objected to the noisiness and the blatant insistence on the naked money —a defect of a merit which most bridge players will also, and the more, object to. But at the race track the real bridge players are apt to be comfortable only in the clubhouse boxes, the one part of a race track I do not like; and they will be offended if not horrified at the atmosphere of a "poolroom" (where bets are, il-legally, made and paid far from the track), most of which I like. The smoke is thick and hot, but I rejoice in the forthrightness and lack of social front it enfolds. The "boardman" pronounces weirdly and has a musically shocking voice but his singsong calling the line is music to my ear: "Tuwenty, tuwenty, tuwelve, naiyun to faive, 'n scuratch the bottom."

And bridge has more than a tradition, it has a whole retinue of observances, case-hardened paraphernalia of doings which seem even more important and inviolable than the integral game. Even in its language bridge must tilt its nose. It is no longer enough to shuffle the cards at a bridge table, they must be "made."

And how they must be made! Shuffling is of as little importance in bridge as it can be in any card game: all the cards are dealt, one at a time, roundabout; and if the tricks are taken and the deck assembled with any sort of honest carelessness it is impossible to tell by scrutiny of the hands whether shuffling has been indulged in or not. Yet bridge will certainly assay a higher load of shuffling than any two other games. The very technique of shuffling in bridge is bridgy—that method introduced by bridge players and practised secretly and displayed with infernal repetitiousness by bridge players, wherein the cards are held aloft, faces exposed, separated, bent in the middle, and exploded together with a long-drawn crackling. For some years while already addicted to other card games I declined bridge but sat in the room through many games, and I well remember the monotonous impression: a period of solemn silence, another period dominated by that machine-gun shuffling over and over while four voices rose in chorus. And just one person, not the dealer, must do the shuffling of the proper one of the two, no more and no less, decks. And you must always deal the "right" deck. The first occasion on which a man half sprang up as I had begun to deal, put his hand over the deck, and gasped "You deal the blue," I for a moment did not know what he was talking about, then promptly lost respect for one I had thought a pretty good sort of intelligent person. Since then I have learned just to say to myself "It's bridge."

Then comes the cut; another fundamental bit of cards which is of minimal importance in bridge. In a four-hand full-deck game the single cut used in bridge can do no more than switch the same hands about the table and one-fourth of the time does not even do that. In other games any player may ask for a cut; in bridge no player ever asks but one particular player always does cut.

All this frippery ritual can provide one of the sorts of delight possible at bridge: you may call for three decks, shuffle out of turn, deal the wrong cards or without shuffling, decline to cut or call for the right to cut. I recommend taking quickly (you must be quick) the deck just used and dealing without a shuffle—it makes the least real difference and causes the greatest consternation. But the pleasure is not a worthy one; and it cannot last long, for either your companions are primarily card players and will willingly put up with you or they are primarily bridge players and will heave you out.

I repeat that my irritability on these scores is *prima facie* my own irritability, and that these tight observances are themselves, by the very force of the charge, not integral to the game itself and are largely just the counter-prejudices of most, not of all, of those who play it. But is it not symptomatic of something constitutional in the game that it should have attracted such a mass of special ritualism and should attract especially such players? I think it is so symptomatic; that the game is essentially pedestrian, formalistic, suited to be eminently respectable, exigent of niceties of procedure but not greatly exigent either of the craft and courage of the gambler or of the insight and consecution of the thinker.

Bridge is not a gambling game, we are told by Mr. Culbertson and the grand chorus. The reply simply is that bridge is, of course, a gambling game but a mild one. Anything is a gamble, upon the broad definition, which risks anything of value on a chance. Upon a narrower and more proper definition (such as will rule out for instance insurance, which in all simplicity is the making and taking of a bet), anything is a gamble which risks something of value which need not otherwise be risked or which is not regarded as an offset to something already risked. Chance cannot in any way be avoided; the gambler takes thought to increase his unavoidable chances, the anti-gambler takes thought to decrease them. Where the gambling chance is erected by arbitrary rules as to the performance of selected objects, the gamble is a gambling game. By modern usage gambling games are restricted to those in which the value risked is money or something directly convertible into money. With all strictness bridge is a gambling game and its player when he plays is gambling.

The arguments offered in support of the frequent assertion that bridge is not gambling are two. Bridge is not dependent upon high stakes. Bridge is only negligibly a game of chance or luck. The second of these statements is I think quite untrue; but the first is quite true. That bridge calls for some stakes, a survey of the almost universal habit and insistence of players is sufficient evidence; but that it does not call for high stakes —i.e. high relatively to the means of the players—the same survey will as sufficiently show.

And it lacks most noticeably that tendency of games generally to push up the stakes among players continuing to play together either for an evening or through a longer period. Crap, banker and broker, red dog will skyrocket in an hour; poker will make the same push felt, surely, in a few sittings. The stakes at bridge may even tend to decrease. These characteristics are shared by some other card games but by none certainly in the same measure with whist or bridge. Where the level of play does increase with an individual or a party, it will be found explained not in the impetus of the game so often, as in increased wealth of the players or in the professional persuasion of one player that he is better than those with whom he plans to play. To be sure, those who love to gamble may also like to play bridge, and playing bridge they may be willing to increase the stakes beyond the particular wishes of others—and yet be content to play for a winning or losing they would not bother to struggle for at poker.

But all this is to say what I started by saying; bridge is

a mild gamble, as a gamble "not much." It is a gambling game in which the gambling interest is essential but not dominant. But if it is not dominant, if it is even a junior partner, it is far from being unimportant in the combination. Without it bridge would be scarcely as good as solitaire and profoundly inferior to checkers, not to mention chess.

This is mostly because of the determinative part played in the game by luck. Many types of solitaire give their successful player more reason to congratulate his skill than does success at a bridge sitting. The nature, extent, and conclusiveness of the "science" and skill of bridge will be glanced at presently; but the frequent assertion that luck is negligible in bridge is silly. The power of the cards dealt, the "distribution," the fitting of the two hands, the placing of the opposing cards and of the opposing hands, the succession of hands and of deals, the toss-up choices of leads, these and how much else are the obvious province of chance?

The upshot of most of the talk of those who decry the luck of bridge is to produce the "law of averages" and aver that "luck evens up." This is as true of roulette or cutting cards. But in those, it may be said, skill has no part, while in bridge if luck evens up the skill of the players remains decisive. Then all that is just as true of poker, of fan-tan, rum, pitch, and dozens of other games which never try to divorce themselves from chance. The half-knowledge of the theory of probabilities on the part of the many bridge players who complacently and compliantly deny luck in bridge because "breaks are even in the long run" is more misleading than no knowledge. Yes, chances come even in the long run. But how long? And for any nameable run no matter how long, if it is probable that there will be many instances of even distribution, it is just as probable there will be some instances of uneven and of blank distribution.

If I toss up two pennies they will probably fall head and tail rather than both heads or both tails. If I toss a hundred pennies and they fall all heads I would be surprised, but not as surprised as I would be if I were to toss those hundred pennies a hundred times and they were to fall every time fifty heads, fifty tails. Point me out at random one man in a gathering of a thousand bridge players, and if I had to bet on his luck during his next ten rubbers at the game I would bet it would be even; point out the whole gathering and I will bet that among them is at least one whose luck will be consistently bad. And somewhere in this world of bridge players there may well be some devoted unfortunate who dies at an advanced age after a solid lifetime of bad bridge luck.

I have seen a roulette wheel come red fifty-three times in succession. And I have sat down at bridge practically every evening of one long summer from June into September, playing with a stable group, most of them not supposed to be my superiors at the game, pivoting partners haphazardly, and never once got up winner. But there is no need to labor the point. Luck does even up for most bridge players over a short period —rarely indeed more than a few years. And their skill remains. But their skill remains of so little final importance that unless they are playing "out of their class" their success or failure will even up almost in the same degree as does their luck.

Let me not argue myself into denying the skill of bridge playing—its reality, its intricacy, its joy. Primarily, in the play of the cards, the skill called for is like that of chess, the recognition of the situation, foresight of possibilities, and arranging of combinations. It is paltry—small, restricted, and brief—compared to that of chess, and susceptible of much less precision and definiteness, but it is of the same sort: analytic grasp and imaginative planning. There is the inference from cards and plays seen to the probabilities of cards unseen. Then there is the element of a highly special kind of card memory, the knowledge of what cards have fallen and what cards are left and where.

This mnemonic is where I am weakest, and I get little pleasure out of it (except in unusual situations) even when I do it with more than usual dutifulness. I know more natural players who seem to have always done it by natural gift and without effort, and others who by training have come to do it without effort; and both probably with little specific pleasure in it. In the bidding, which whist in its modern forms has more and more made prominent, there is the exercise of judgment in valuing one's hand and, with the progress of the bidding, in more exactly valuing it with reference to the others around the table, and the exploration and "bluff" of competitive bidding—a natural fascination except in those sad sessions when one never has any cards to value. This is about all. Surely, it is much.

And yet in the view and usage of most bridge players these exercises of skill are not all. These integral grapplings with the nature of cards and the laws of the game (except the card memory) may even drop almost out of sight and out of existence under the predominance of another factor which clutters up our bridge tables: the mass of "rules" and "conventions" which the poor conscientious bridge player, without the conscience of a card player, learns by heart and repeats by card. As an ideal herein the player approaches the state where he need never think but merely remember—and then prate of bridge as a triumph of skill. It is perhaps what is meant by the still more fond characterization of it as a science.

In this extremity the game is reduced to a laborious method of cutting cards for small stakes, plus the pleasure, if there be any, of repeating nonsense verses. And this extremity is not merely a distant limit, it is one often closely approached if not accomplished. I have known "book players" at chess, but they were feeble and besides constantly left beyond book help; for the complexity of chess can barely be scratched by all the indefatigable analysis which has been devoted to it— interesting, sound, and useful as that is. But it seems possible to rig up for bridge a roughly adequate "book of conduct": hideous in its prolixity, more hideous in its arbitrariness and creakiness. It will not make its slave a very good player, but it will not often leave him altogether at a loss. It is at once the perfection and the *reductio ad absurdum* of pedantry.

The worst of this bridge pedantry lies in the nature of the largest factor: the "conventions," which have no reason or purpose aside from their arbitrarily concocted "signal" value. Obviously it is good to know the faces of those cards the backs only of which can be seen. If I flash my cards to my partner in a mirror or pass him a note under the table, that is cheating. And if I prearrange to let him know them by means of my management of cigarette or of voice in bidding, that is cheating. But if I prearrange to let him know by *what* I do in any of the regular doings of the game, that is not cheating but the highest-class bridge.

I quite agree such a recognized practice is not cheating; the line between it and what remains cheating is clear. It is not cheating the opponents; but it is cheating the game. Of course, a great part of the proper pleasure of the game is in inferring from a partner's or an opponent's bid or play his holding: knowing the game, observing his action, the situation from his point of view is more or less accurately to be guessed. But when a play carries a meaning, which could not possibly be inferred by any acuteness of insight into the game, then I call that prearranged signal illicit; not illegal but extra-legal, a threat to the integrity of the game and a weariness of the flesh.

There is that pet abomination "the echo," the use of which I read in the paper last night is the "necessary sign that one is no longer a dub." There are the discards chosen not from any plan of one's hand but to tell one's partner what to lead. Yet there are bridge players unwilling to see any difference between the most natural inference-allowing plays and the most arbitrary signals —probably because they have taken them all from rules learned by rote. It was once said to me when I complained of conventions: "But you use them—you led queen from queen-jack-ten—why not jack or ten?"

It is in bidding, however, that the convention evil has really taken root and flourished, until today it is an unwary player who easily supposes any bid means what it says. Here as in the play of the cards there is licit inference and room for trickery. In auction I have bid a blank suit over a no-trump bid to my right in hopes of sending play into a suit where I might do some damage. Honest trickery is good game but the far opposite of convention: the one hopes to be believed, the other expects to be understood in a crooked sense. So today at contract my partner bids four hearts and without a diamond in my hand I am supposed to bid five diamonds; not that I mean to bid anything at all but just to tell my partner that I have no diamonds! It is my contention that that game which cannot successfully be played by intelligence working on a knowledge of the laws of the game and of the nature of the objects involved is to that extent a bad game. If bridge requires the paraphernalia of arbitrary convention which now encrusts it, then to that extent bridge is a bad game.

I do not believe bridge requires it (I am not so sure about contract), but there is a characteristic tendency toward it. Inference-allowing plays run vaguely into conventions. I can imagine situations in which what is called an informatory double might be tried by one who had never heard of such a thing. To lead ace-deuce with no more of the suit in hand is often a natural lead and readable by one's partner. The obvious lead of king from ace-king might easily be shifted to ace when one holds no more of the suit and means to lead the two tops quickly. So the more natural conventions arise and the arrant and unclean ones follow. But if this affords some excuse for the manufacturers of conventions, it shifts the complaint to the game, which essentially pushes itself on, unless constantly guarded, to illegitimacy and sterility. And when a game reaches the state in which some players now have bridge, it is ready for suicide—or ought to be were it not that all of us sometimes and some of us always actually like to wrap the living in the trappings of death.

Indeed, these latter complaints of mine are not a detriment but an advantage for two classes of devotee, two classes large and important today in the game's swelling front. There are bridge instructors who are now finding in bridge a "good thing." I have no acquaintance among them, but I believe many of them are good card players who cannot be blamed much for what is in part simply making the most of a gorgeous professional opportunity and in part the succumbing to what all teachers have more or less to succumb to. At best to teach a science or an art is difficult and often useless. Not so with a catalogue. To teach bridge "as she is" is a procedure sufficiently long but not of necessity too long; cumulative but with stopovers anywhere, with practice provided and attractive, and with the subject teachable with or without card-sense—teachable as the Chinese alphabet. The rest of the same picture is the number of docile pupils for this lore: those who are not especially card players or (without any prejudice to them) keen for games anyway. Bridge is today "the thing"; they want to play it respectably, as little unprofitably as may be, and particularly they want social ease in the performance and the lingo. And surely for all this the conventionalized and rule-ridden game of bridge, especially contract bridge, is manna.

Contract, the most recent and ingenious evolvement of the professionals, seems not to be merely reprehensible. Its scoring, if not perfect, is in some respects certainly more rational than that of auction, and it has undoubtedly added elements of freshness and probably of permanent vivacity to the game. Yet it has curiously added to the frightfulness of the language; and, what is fundamental, it has seized on that worst aspect of bridge, the regimented and conventionalized and hypocritic, for emphasis and development—bringing whist to that peak of fashionableness but also of absurdity where it now finds itself. Not long ago I silently watched an eager instructor, volunteer-instructor, who had just had a brief course of instruction in contract, as he was revealing its mysteries to a half-dozen eager friends who played auction but had only heard of the new game. The master was a man of at least some academic standing, but it was hard to believe that a mind of any scientific habit could be seduced even by bridge to such a farrago. Fresh hands would be dealt face up and on them directed a stream of imperatives as to what each must successively bid and do. And the others, the learners—no one of them ever seemed to think to ask why any of these strange new "musts" should come from what newness there was in contract, or for that matter why any of the "musts" were anyway. They seemed to think the change of law which made auction into contract should be left unquestioned in the code and to accept the new game as simply something to be "learnt" only by accepting a multitude of prescriptions from someone who had himself had them from some other one presumably just come down from Mount Sinai.

I am confident I could take two good card players who knew auction but had never heard of contract, confine instruction to one sentence: "Contract is auction in which you score toward game only the tricks you have bid, further tricks scoring as honors"; and without any consultation between them set them down against any of that group with all their instructions and signals; and with any break in cards my pair would in twenty rubbers clean up. Later on they might not—after the instructed ones perfected their instructions. I would be tempted still to bet on the signalless card players; but it may be that contract cannot be successfully played without some scaffolding of conventional "system." In that case it is just a bad game.

Of all games which combine luck and skill, draw poker is incontestably king. The luck is raw and strong and rapid in its repetitious decisiveness, with just enough interval between threat and catastrophe. The powerfulness of luck is nicely met by equal vigor in the factor of skill. The player considers a complexity—of card values, of probabilities of the draw, of position play, of ratios with the pot, of capital, of psychological habits and twists, of surface indications—which all the more restricted intricacy of bridge can match only in intricacy. And he must often do it in the lifting of an eye. He needs beyond the fenced-off analysis of the bridge player both the patience and the swiftness of a cat, and the intellectual vigilance of a trial lawyer.

One of the most wearying of games it is, beyond all rival, the hardest game to stop. I have played it forty hours at one session and been willing for more. I have seen it begun "for an hour at most" and go over to breakfast; trains missed, engagements ignored, wives lied to, the world well lost. It has intoxicated me often, bridge never. Beside it bridge seems a languid thing.

Yet I scarcely ever play poker; I often play bridge. And when I play poker it is apt to degenerate after a while to stud, to seven-card stud, to wild cards, to dealer's choice, to all manner of more pure and puerile gambles. Poker is just too good. I no longer care to work so hard as good poker requires—even if I could by working so hard recapture that greater joy than bridge ever knew, which such hard work once gave me. And I have learned that one (at least almost anyone) does not play good poker unless he plays for more than he can afford to lose. I am no longer so often glad to play cards for more than I can afford to lose; and unless you at least try to play good poker it, too, is only a laborious method of cutting cards.

Why is this marked difference as to stakes between poker and bridge? In part it lies in the harder work poker demands. Not only does one expend more energy in playing his best poker than in playing his best bridge, but also a partial relaxation of that energy is not so fatal to bridge as to poker. In part, the reason lies in the fact that in poker the skill is closely wedded to the gambling: poker is more directly a battle with the players for the stakes, bridge more directly a battle with the cards for tricks. And this suggests a more subtle third reason. The accomplishments of poker are fragmentary and mostly secret (and not particularly interesting) except to the player himself: without the stakes to show for it he grows weary of at best self-commendation. The accomplishments of bridge are more extended—brief campaigns but still campaigns of some extent and some coherent unity—and they are displayed and public to the table each side of which has been involved.

In this merit (and as a merit I recognize it) bridge approaches chess, the incontestable king of games of pure skill. But how feeble, in the accomplishment as in the skill called for, is that approach! I played chess before either poker or bridge, and with more voluntary enthusiasm and determination; but I gave it up more completely. It was not a deliberate giving up, but I have known prudent men with whom it was. I play it now, but too seldom for full commitment to its embrace. And it may be said in its favor that half-chess can be played, as half-poker cannot be played, with considerable pleasure, without loss of the quality of its charm.

Stakes are quite impertinent to chess; it is indeed better played with none; the game is all. And the game is rather too much, except for those born to it; too hard for mind and also for nerve. I have sounded ridiculous to many a college athlete, full of glory and cheering-sections, by saying chess is the most exciting of all games; but I am persuaded this is true. I have come out of chess matches limper than any dishrag and gone home to a sleep tormented by unceasing chess situations (which, I believe, I have often handled better, piece-meal fashion, in my sleep than awake). So, too, I have violated sleep with cards after hectic sessions at poker—or red dog or banker and broker. And once after years with no thought of chess I came from a poker battle to find in sleep the old obsession of chess. But bridge has seldom or never thus against my will perpetuated itself upon that inward eye which may be the bane of solitude. Let us then be thankful for the mildness and pleasantness of bridge, for its merits but especially for the merits of its defects. Inferior to poker and to chess it does household tasks for which their strength unfits them.

I am not tempted to argue the swelling claims made for it as moral and intellectual training ground. Let such claims bemuse the zealots and justify the careful ones who seek justification for their pastimes. No doubt any practice in the habit of concentration on the matter in hand is some help toward the advantages, and disadvantages, of that habit; but the transfer of special abilities, even when they are abilities of general faculties, is slippery ground. I once saw a man finger through a deck of cards as rapidly as they could be read, then deal seven poker hands face down one card at a time, and call off the combinations in each hand. I would avoid playing poker with that man, but I would not be confident of his showing unusual ability in other lines even similar. I have played blindfold chess; I never found that performance any gift-bringer to my memory in other fields. Every game, being a competition, has some disciplinary force; being a technique, has some intellectual incitement; any game specifically has precious little.

But of the worthiness of bridge as recreation I have the highest opinion. It "cheers but not inebriates." It is a gamble, poor just as a gamble, at which we can enjoy a mild excitement which will not exhaust, and find a not too severe relief from the greater and more unavoidable gambles which crowd most of our days. It is a test of intellectual skill, poor just as intellection, in which we can exercise our judgment and ingenuity with a mild concentration which will not exhaust, and find a not too severe relief from the intellectual void into which most of our days are crowded.

But in its mild hospitality is a sting. We need only survey the tables drawn out after lunch or even break-fast by mountain and shore and thronged until or through the time for the dance, to realize how much it may be an anodyne, or an engulfer of hours during at least some of which the bridge table is only an escape from more positive, and more exacting, opportunities. And when its mildness comes toward its limit, when it is deadened by all that can make bridge deadly, then, apart at least from its social or its professional aspect, it sinks through pastime into the merest way of passing time, and then of killing time. There is no longer question of good game or bad but, my masters, of sin. All of us kill time at times, but we should not ever; for it is simply suicide, partial suicide. Emotions which, like most human emotions, waste upon illusions the hope, the joy, the fear, the grief due to reality were held by the Stoics to be the "conditions of sin"; but the boredom which surpasses and caricatures the apathy of the Stoic and concentrates upon the merry phantasms of cards the listless blankness it has already drawn over reality would outrage the pious and logical Stoic more than hilarity or vexation. I call on the Stoic because he is the supposed admirer of apathy. To kill time is to serve the devil; not Lucifer who was Son of the Morning, but that older more abysmal devil from whom even Chaos and Old Night were young brave rebels.

The coupling of "disputes" with "whisk" in David Hume's question may seem curious to us. The disputes in which bridge is tritely plentiful are evidently not Hume's meaning; and disputes in his sense are more apt today to be among the things against which bridge serves society as a guard—the more real perspectives of dispute or even conversation. But we must remember that in the quotation we are among the "scavants" of mid-eighteenth-century Paris, when "disputes" were as much the fashion as "whisk." And in some circles, for bridge has many many circles now, that particular division and congress of interests would still hold good.

And it may also seem curious that it should be Horace Walpole who was the malignant toward whist. For, though the characterization set for him by Macaulay is unfair, he is the sort of person we should expect to find at least an occasional addict. It was not that he was without interest in games (a defect to which all types of character are susceptible), for he went gladly to play at "loo" with my Lady Suffolk or the Princess Amelia. I think the sufficient reasons were his antagonism for the "scavants" and other whisk players (still a potent influence), and the fact that he was essentially a humorist. Bridge even in its ancestral forms has always been peculiarly "vulnerable" to humor.

I am not a humorist. And so from this trifling solemnness I will respite myself in the masterpiece of solemn trifling, bridge—will go forth hoping to be asked to "make a fourth," enjoy playing as good bridge as I can play, missing or ignoring my partner's signals and dealing with unshuffled decks, and come home mildly elated that my skill has won me a couple of dollars or mildly chagrined that my bad luck has lost me a couple.

8

A Defense
of Horse Racing

The Voice of the people is the voice of Hearst. One day among the sentiments supplied by his satellitic writers I was halted by a moral essay on horse racing. The author declared that "some people have low tastes and others have refined tastes," going on into a monograph on playing the races as a prime example of low tastes.

Here we have a

. . . vulgar amusement because it brings the player into contact with low and, in most cases, vicious people; it must be done surreptitiously, its purpose is to gain something on a gamble, the races are not always honest and the whole atmosphere of the transaction is usually sordid and unclean. Those who "play the races" regularly have low taste.

And this gamble is worse than other gambles. For you "know by instinct" that card games are in better taste than the races. Not only is the poker-player estimable in comparison but so also is the stock-gambler, "the Wall street magnet who risks millions in a stock deal." This is true especially since, "if both pursuits are immoral, the Wall street immorality does not bring one into contact with such terrible people as those who are interested in horse racing."

The day after the appearance of this essay the same paper in an editorial expressed the hope as almost too good to be hoped "that the police possibly have been made aware of the disturbance of business and the occasional defalcations caused by gambling on horse races," and that "we may, conceivably, be on the way of getting rid of the whole brood."

It would not be easy to find passages more tempting to criticism, and, even among the rapid wisdoms of daily syndication, more open to criticism, logical, factual, implicative. But my interest here is apologetic and encomiastic—not exegetically critical; the quotation a text not an object. For I am one of "those who play the races regularly"; and though I am not moved to attempt refutation of my "low tastes," I am moved to a bit of comment on the races, playing the races, and some of the philosophic suggestions of those activities and of the defenses thereof as well as the attacks upon them.

Now I am just a college teacher and not a "Wall street magnate who risks millions." Indeed I confess I have never played the market and have had a sort of prejudice against the market as compared with the track, as a possible game for me, because it takes too much money; and as a game considered from the theoretical view of ethics because it does not display the variety, the color, and because it involves an element both of economic mischief and of hypocrisy in gambling with what is not primarily a game, namely the operating capital and consumption goods of the world. It is doubtless a good game, but its pawns have other uses and importance; and as a game it is not as good as racing.

It will be seen that my philosophic defense takes racing as a game. Those who find therein sufficient damnation unless it can be shown the game is needed for future projects may dismiss the present defense forthwith. My thesis is that the life of an actual man in this world is frequently, not always, better if it includes playing the races, and better irrespective of whether he thereby increases his or anyone's wealth, position or material comfort; that this world is better in the eye of God with racing in it than without, and so not because God therein foresees more money in the banks tomorrow but because what he sees is pleasanter.

Racing is actual and therefore not immaculate. It has been worse. It may be made worse. Some real trepidation there might be lest the races become one of those good things made bad by attitude—as the saloon in too many instances became what its enemies said it was, as witchcraft may have become what Christianity reviled it as being, and as how many lesser things and perhaps other greater and more essential and more gracious things than drink have been contaminated by opinionatively attributed shame.

If our Hearst author were by some curious fatality to find himself playing the races, he would do so with a social shame and a troubled conscience; there would be reticence if not subterfuge and even the temptation to lie; incertitude or hypocrisy might creep into the limpidity of his daily exhortation; from it all might develop bravado or shamelessness or hopelessness of salvation or permanent disease of conscience. I know of a man, respectable successful businessman he was, who went one afternoon to Pimlico but was miserable for fear he would be seen, especially be seen making his bet. Fortunately he had the social and business judgment not to continue in the way, and although a little later he tricked his employees and his bank in an ingeniously crooked device he did it in a large, almost Wall-Street, manner and apparently was not ashamed.

Now our columnist does not mention the official justification of racing—the improvement of the breed. And though there is no particular reason why he should, it is a pity he did not; for then we should have had the pitifullest aspect of the defense along with the illogic and falsity of the attack. Not that I deny what the protagonists of racing always urge on this subject. It is true. The testimony of military authorities is conclusive that if you want the army to have the best supply of horses you need the thoroughbred, and the thoroughbred is the product of racing. And the services of the thoroughbred and of current racing to the farmers is undeniable. Yet this argument is not only insincere (though not dishonest) since it expounds what is not the actual motive of those who for various motives maintain racing, but, at best, strikes one like the argument that urges the future lawyer or physician to spend years in Latin grammar and reading Homer so that he may be able to use the tags of Latin and Greek that still officiate in law and medicine.

And yet I think that, differently considered, the improvement of the breed is, surely not the only, but a large and basic and proper value justifying racing and not to be achieved otherwise. And why? Not because it speeds the plow; but simply because the thoroughbred is one of the most effective and beautiful things in the world. And I think if those who honestly support racing were to pass beyond their desperate military and agricultural utilities and speak of the joy which comes or may come from the thoroughbred horse, distinguishable from and additional to the joy of the race or of the gamble, they would have a juster argument, though perhaps not so weighty with legislatures.

And if the thoroughbred has a value in himself, he adds to that value in the race. For this came he into the world; for this his breed has been improved, and here he finds and shows the Aristotelian virtue, the worthy performance of one's own activity, the function of one's essence.

And in his virtue is his joy and a joy of the beholder. The bear may object to his baiting or the bull to the ring; the cock's spurs are an episode and a pervertible instrument; to the human gladiator his contest is a profession; the thoroughbred *is* a racer and, except for the seldom one which by bad nature, bad treatment, or bad fortune becomes sour, he loves to race. Did you watch June Flower get away from the gate during her brief unbeaten career? Did you see Exterminator the day he won the Philadelphia Handicap at Havre de Grace? After six years of arduous campaigning against the best over all tracks and with high weights, he cantered down to the paddock, and when he heard the greeting to him which ran down the grandstand from the eighth pole to the clubhouse he turned and danced like a two-year-old. And then he came back from the paddock in parade and to the post the calm concentrated champion, oblivious of the crowd, watchful of the starter and the tapes, patient but wary of the impatient youngsters he was to beat.

Is it disputable that thoroughbreds running make the best of races? Is it contemptible that the race is perhaps men's oldest dramatic interest? And if the stage as the figure for human life is a commonplace, so too is the racecourse, which in Thomas Hobbes is developed as a figure with splendid rhetoric and becomes more than a figure. At the racecourse is not artful simulation of humanity; in this is its great lack of the drama's richness and possibility; but in its defect is virtue of purity and authenticity. Here, too, is relief from self; joy of spectacle; endeavor made objective; catharsis by fear and pity; not only in shape and color and contest, spectacle and comedy; but also in achievement and frustration and tragedy.

Were you at Pimlico when Exterminator was defeated for his third Cup? Two miles and a quarter, the longest race of the year. Johnson is down with appendicitis; who will ride Exterminator? A little apprentice boy is put up; he has a string of successes at the meeting; this is the climax and the proudest moment of his life. But isn't it foolish to trust him with a champion and in a cup race? The Big Hoss knows, he knows more than most jockeys anyway; he has won for many different riders. But one thing he cannot know; how far he is expected to travel; he has won this season at all distances from six furlongs up. Off they go and away at a suicidal pace. This apprentice's only idea is to get to the front and stay there. To do this he must outrun Exodus, a crazy sprinter with a feather on his back. A half-mile, mile, mile and a half, Exterminator on the outside, head and head. Even I in the stand without a watch, I who never rode a race and have no judgment of pace, see what is happening. A mile and five eighths; Exterminator raises and drops his tail once, he stumbles and falters momentarily, then strides on, steadily and with the rhythm of perfected technique, but much more slowly and with evident painfulness. The heavy weight and the great distance today, the heavy weights and distances of unresting years behind, focused in the frantic driving pace of this two-minutes, have finally conquered even his iron legs. But he has already put Exodus away. Exodus swerves, wobbles, and as his rider tries to pull him up goes quite crazy, bolts to the outside fence, staggers almost backward on the far turn. Captain Alcock and Paul Jones who have been kept far out of it by decent riding are now moving up. In the stretch they pass Exterminator still grimly drawing to the goal which is just beyond his leadership.

And did you see Billy Kelly beat Sir Barton and Mad Hatter in the fall of 1920? It was, I think, the finest race Sir Barton ever ran, a monument to the courage of the great thoroughbred and to the greatness of courage, though the time was slow.

In addition to the horses and their race as immediate perception are all the associations of each, the traditions of the race if it be a "classic," the personality, the history, the ancestry of the horses, and beyond these the jockeys, the trainers, the owners, all the human and institutional background. An erudite and acute professor once said to me that the race track was the most tiresome and uninteresting of places of supposed entertainment, that no one not fevered with gambling could go to enjoy it, that he had been there once. Well, there is such a thing as willingness of imagination, as knowing what's what. Packed in the moments between bugle and red board there is a wealth of experiential incident to fill a professor's course. And between races ("after one race" my professor said "one just waits half an hour to see

another"); before and after and between races, there is
the air and the spectacle, the faces and colors of the
jockey-house, the paddock and the variously individual
horses there, the smells and the sounds, the consulta-
tions in the stalls, the saddling, the instructions to the
riders, the paddock call and the call to the post; and
always the crowd with its types and peculiars, touts and
come-ons, veterans and new enthusiasts and casuals,
handicappers and system players and followers of "information," the crowd with its wisdom and its superstition and its veering fashions in opinion, its amazing
shrewdness and amazing human sheepishness.

All this aside from betting. The two most memorable
and remembered races I have seen were without financial
interest to me. And a number of times I have got together enough cash to take me to the track when I could
not get wherewithal to bet. But I have no desire to leave
the betting out. It is not merely that it heightens all that
I have spoken of, heightens its color and intensity as
present experience and that it is indeed to the betting
we owe the spurring on of our lazy faculties to the
astonishing knowledge we have about horses and races;
gambling may be worth while in its own right, and the
race track offers the finest of all gambles.

Just why is gambling looked upon so darkly? I know,
of course, that it seems generally supposed that "money
is gambled *away*." So our elder advisers tell us, and so
the newspapers imply when they recount after each de-
falcation that the errant cashier played the thousands or

hundreds of thousands on the races. The only logical solution of this cashier anomaly is either that for every reported case there are ten or twenty cashiers playing someone else's money on the races and not losing so extraordinarily or else that there is something peculiar about a cashier which makes him able to pick so appallingly many losers.

The money that goes into a bet must come out somewhere not as blue smoke. To be sure when we play in a game which is cut, it behooves us to be aware thereof; and the races are cut. But the cut of the mutuels is very small (and should be smaller), and we should not object to paying for the upkeep of the institution we use. But in the end there are, I think, just two real objections to gambling (and I do not deny that they are legitimate but do deny they are conclusive): that it uses time but is nonproductive; that it is an intoxication and dangerous.

We need not dwell upon the extreme importance given to accomplishment, and especially economic accomplishment. We may admit the good of productive accomplishment and we may admit that pure gambling, though it almost as frequently as not increases the individual's money, does so only by taking from someone else; it creates no capital, and hence from a social point of view is waste of capital or of time that might have produced capital.

But just so far the gambling-place might claim to be little, if any, worse off than many another more repu-

table place—than the theater or the tennis court or even
the grocery store. The trouble is that from the productive point of view gambling has peculiar defects or
dangers as compared with the other places. When one
goes to the theater there is on the one hand some check
upon continuance in growing tired (though theater-
going itself easily becomes a habit-drug) and on the
other hand there is no delusion but that the expense is
definite and steady. Gambling carries the hope and possibility of paying its own way or more, thus tending to set
itself up as itself instrumental to direct money-getting
that is at the same time not capital-making; and its in-
toxicating power is much more direct and constant. The
first of these aspects of gambling, although it validates
the social distrust of its waste of time, is in itself for the
individual a good, and it probably makes gambling,
considered just quantitatively and despite the violent
superstition to the contrary, the cheapest of all the
world's amusements. Note that to the basic charge that
playing the races is gambling and hence nonproductive
the reply is that all goodness in the end is nonproductive;
that doubtless much time is spent over form-sheets
and before mutuel windows which might be better spent
but that the same is true of sleeping, eating, listening to
lectures, and reading the Bible; that all those employed
in carrying on racing, from the newspaper selectors to
the judges in the stand, are productively employed in
the same sense that the philosopher and the contortionist
and the farmer are: that of those who only play the

races the one real and complete case is against the professional gambler who plays the races and does nothing else, and that even this almost mythical personage is not to be too harshly judged even from the merely social point of view, since no one knows how bad a preacher or lawyer is spared the world in his being a gambler.

The question of intoxication, inebriation, ecstasy, whatever word may be chosen, is too large to be dealt with here. But it is at least safe to say that the case against it has never been sufficiently made out, on the utilitarian grounds that seem the chief modern basis for its derogation or on any other, to justify complete taboo. And if we are prepared to oust it root and branch, then art and religion at least must go with racing and drink. I would not maintain that the possession of an intoxicative aspect is sufficient justification of anything. I am not a morphia-addict. For one thing it would interfere wth my playing the races. But I am persuaded that the fact that race track gambling may intoxicate is not condemnation but merit, and that among intoxications racing may hold up its head. There is a certain surreptitiousness, though surprisingly little, about it, and the badness which goes with that: but herein the fault is of the laws, not of the races.

Few persons place bets who have not sometimes been in the open-air magic and imaginatively lingering atmosphere of the track, almost none who does not have some knowledge of the actual animals and more-than-monetary interest in their performances. The conductor

and the waiter who leave their fifty-cent bets with a
runner, the dentist and the lawyer and the merchant
who call in over the phone at lunchtime, even the packer
who takes a bit of his employer's sacred time to find out
the fate of his hunch or his tip or his handicap selection
—are they all in the grip of a fatal and debasing de-
bauch?

They swear at the horses and that not seldom. There
is nothing more intricately, perfidiously, and inexhaus-
tibly exasperating. But there are many who have no
profits to show who are more than secretly grateful, and
many more who should be. I came out from the first
day's racing at Bowie in April and on the train took a
seat beside an old gentleman who presently burst into
speech. He had gone broke after the fifth race. For
forty-one years he had played the races regularly every
spring. For forty-one years he had lost. That was not
so bad; he always hoped he might win, but he did not
really expect it. Every year he worked through the
winter and put aside a little capital, then started in the
spring and played as long as he could. Usually he could
last almost through the season; but he was great on
doubling up on his losses and sometimes he would play
it all and lose. But this year he had gone broke the
first day—before the end of the first day. That did hurt
his feelings, and he had come out to sit in gloom in the
train. Yes, it is very well that most of us are not so extreme in our devotion.

Doubtless these conductors and waiters and lawyers

and teachers and bootleggers and housewives and clerks would do better to read Shelley or listen to Bach or meditate God from the top of the mountain or by the sea. Some of them do at times. But there is some advantage in availability, some in variety, some in artificiality, some even in a certain crudeness of intoxication. And not all are ready for the heights nor are the heights perhaps always best.

The defalcations and the absconding cashiers, of which we hear so much, are a sort of vivid combination of the two general arguments: the positive injury to economic life and to social morality through individual allurement by the races. And this I should be disposed to grant. I doubt most of the specific instances and the universal assertions, but I should be grieved to doubt the power of the glory of race track gambling. Such dangerousness is almost the measure of a thing's value. For all good things must be dangerous in a world where all things may be used as instruments and where the keener the edge the deeper the wound. Socrates was fond of saying that the physician is best not only at saving but at taking life. When the churchman is shocked by the Epicurean eloquence of Lucretius's line "To such damned deeds religion urges men," it may be pointed out that therein is tribute to the greatest and most abiding intoxication man has known. And if men steal to play the races they also steal for wives and children and women who are not wives. To be sure, it has been proposed that we abolish wives and children. But then women also steal for men.

And now it may be said we are left with the "terrible people" with whom one who plays the races must associate. Well, I deny it, and I do not know that much argument is relevant. One curious conjuncture of ideas might be noticed: that between racing and crookedness, racing people and dishonesty. Curious, because where in the world at any time can one find a business of one-hundredth the volume of that which is carried on every day about the races, largely by phone, and without signatures, guarantees, security except mutual honesty? In the sense of personal honesty about money commend me to the racing-man and the gambler. Is that strange? Do we not usually trust him who has been through the fire? That there is some, even considerable, of what is meant by "crookedness" in the actual racing of the horses is doubtless true.

There is an occasional trainer somewhat of the temper of him who muttered at the Fair Grounds in New Orleans, "I'll teach the damn' public to play my horses"; jockeys have been known to ride curious races; several races a year are run the night before. Yet to anyone who for some time has known this most closely scrutinized of sports the astonishing thing is how little crookedness there is and how little it affects the player. There is less than there is in ordinary business. And it need scarcely be said that racing is far more honest than the plots of most movie and magazine stories. There is also the man who goes to the track and is *not* a gambler; who loses a bet he thinks he should have won, and is forever convinced that he was wronged and all races arranged to cheat him. Likewise he who is told to play certain horses, sees them win, and therein finds proof that all races are fixed.

At the track one finds a humanity almost as various as humanity. It is one of racing's charms. Yet there is a singleness and simplicity of direct aim which makes access common without violation of intimacy. At the track I can feel always at home, never intruding, never intruded upon. I can companion carelessly with the darky rubber in the paddock and the owner in the club. And when I am most disgustful of company and resentful of lonesomeness I can find at the track a populous solitude which is neither alienly engrossed like that of Broadway nor personally exacting like that of society. Those who play the races regularly have largely recovered from that careful and fearful respectability which palls so much of society.

"Terrible people?" I have been in academic seminaries and faculty clubs, student organizations, gatherings of the socially proper, or artists, chess-players, athletes, and of business and professional men, even of newspaper men and columnists. And from time to time I have been tempted to feel that each was of "terrible people." But far less often, I think, at the race track than elsewhere. And—perhaps more significant if less logically valuable —from time to time in all of these surroundings I have been tempted to feel that I myself am a "terrible" person; but less often than elsewhere, I am persuaded, at the race track.

9

Idols
of the Twilight

"SPREAD THY CLOSE curtain, love-performing night,"
says Juliet, in the full chorus of lovers, from the Pro-
vencal deploring the dawn to yesterday's beau turning
down the parlor light and to this evening's boy friend
looking for a dark place to park. And on the part of
commentarists: "Wherefore, when we go about to make
or plant a man, do we put out the candle?" says Mr.
Shandy.

Wherefore indeed?

(That we are beginning not to do so is the better
for our question. If we still put out the candle with the
regularity of heretofore, there would be only theoretic
interest in asking why. But if it is already a changing
rule, then there is not only history served and theory
but also the pleasure of wondering what is to come.)

Mr. Shandy, of course, did not ask the question

honestly, but rhetorically; it is an answer not a question. "I know it will be said that in itself and simply taken—like hunger or thirst or sleep—'tis an affair neither good or bad—or shameful or otherwise.—Why then did the delicacy of Diogenes and Plato so recalcitrate against it? and wherefore, when we go about to make or plant a man, do we put out the candle?"

Several secondary reasons for the lover's love of the dark are apparent. It is a measure of safety, from being seen by those who would prevent, gossip, or revenge. It is an initiatory palliative for frightened modesty. It is a cover for imperfections of body or technique. And both as symbol and warder it encourages the positive value of intimacy: night is the time of privacy and sight is a rover.

But obviously the basic reason is Mr. Shandy's: we are well taught that in itself "it" is shameful. Whether it be really so, as Mr. Shandy implies, and the cause of "every evil and disorder in the world of what kind or nature soever, from the first fall of Adam down to my uncle Toby's (inclusive)," or whether it be indifferent, or whether it be good, or whether our holding it evil be sufficient to make it so, are questions beyond; that we are well taught it is shameful is the sufficient reason. We may not altogether believe that it is altogether bad; still we believe its badness is "that which is to be believed."

The official doctrine (Platonic-Pauline-monastic) then is: sensual desire and gratification are evil, especially in the play of two human bodies; virginity is "the better way"; marriage is permissible, and within marriage and for the sake of offspring man and woman may come together, but voluptuousness therein is a yielding to the lusts of the flesh which are evil, and voluptuousness apart from that marriage and apart from that purpose of offspring is sin and perversion.

The doctrine is within itself at least consistent, but its consistency had to give way in practice. The male half of humanity at any rate quickly agreed to leave virginity chiefly to women, to adore and demand it at once and make war on it, to pursue lechery while calling it bad but also good, as something properly frowned on by the church and just as properly required by masculine society. Now putting out the candle, which was strictly consistent with the official doctrine and a great help in fixing it and maintaining it, was also consistent with the masculine view and practice which was itself inconsistent with the official doctrine; and this business of putting out the candle helped greatly to fix and maintain the masculine tradition and of course to safeguard its performances. Thus men's practice continues the atmosphere of that doctrine from which it is a partial and week-day revolt; and putting out the candle is the symbol by which all variations of conventional habit agree and the world, the flesh, and the church live together in sufficient amity throughout Christendom.

That which was chiefly accepted from the official theory by the covertly rebellious masculine practice was the emphasis upon the act itself, the pushing away of its esthetic and personal accompaniments, developments, qualifications, or justifications. For it was the insistent primary hunger of the flesh which forced the breach in ascetic theory, and, this demand satisfied, practice itself would not necessarily push beyond—especially not as long as the candle was always put out. To be sure there was always more than this actually—a man likes some women better than others, partly on grounds general and esthetic, partly on grounds individual and personal. But on the whole it is the femaleness of the partner which is sufficient and the physical release of self which is final; and putting out the candle helps prevent elaboration of procedure and selectivity of object. Lack of light narrows the act by cutting off visual gratification and awareness of the partner's consciousness the while, and by the same token it broadens the promiscuity of choice.

Not all candles have always been put out of course, but those that burned on have been exceptions. Versailles is said to have introduced the horror-raising fashion of uncurtained beds and nearby mirrors, and mirrors would have been scant use without candles; but the fashion scarcely went beyond the pagan circles of the court (and bagnios) and scarcely beyond the pagan period of the high Enlightenment. Though curtains in time went out for good it was for quite other reasons.

Masculine practice is at bottom simply the satisfaction of hunger, but hunger accompanied by human consciousness quickly breeds itself a continuing appetite, and appetites are determined by all sorts of theories and theoretic implications of practices. Masculine practice has been kept pretty close to base by the pressure of the disapproving official theory and by the absence of light. Yet, even so it has had all manner of special trappings with the fashions of periods and has been consistently more continuous and more a source of "vulgar" humor than seems required by the rhythmical seriousness of mere physiology. If now, mainly because of the decay of the official doctrine of asceticism, we stop putting out the candle, practice may easily diverge and alter, invite new theories, and be further changed by the theories which prove fashionable and lasting—by different interpretations of sensuality, different accentings of the orgastic item, different ethical backgrounds, and different formulas of practice derived therefrom.

Prophecy braves the infinite of chance, and all we do here is suggest consideration of the chief possibilities which putting out the light has stood in the way of. There is first the development which putting out the candle has least directly stood in the way of and which, in part for that reason, has gone the farthest and reached the most stability: the this-worldly utilitarian development. This-worldliness has been the most persistent of our tendencies since the Renaissance, but it has come down, under the pressure of business and social success, from the heights of the romantic this-worldliness of the Renaissance, the violent rebel against medieval other-worldliness, to a Main Street this-worldliness which makes its peace with the church on a basis of anti-sensualism, "service," and respectability; an energetic gospel of ambition which leaves ambition small vista beyond money and position, a hedonism which frowns on pleasure, an intellectualism which fears the radicalism of intellect.

With regard to sex its push has properly been toward honesty and naturalism but also toward inadequacy and shallowness; to remove the taboo from sex as a natural fact, but not altogether a nice fact; to take its devilishness from it, but not give it any godliness; to regard it as negligible in importance, though not negligible in fact, to be got out of the way for business and service; to provide against its overemphasis by "sex instruction" or punish its power by social disapproval and cure it by "mental hygiene"; to leave it as a conjugal gesture and a party indulgence. In its conjugal aspect there is thus apt to be skimping and sentimentality, in its party aspect coarseness.

In this development, too, of course there is the separation of the masculine tradition; so that there is a junior as well as a senior doctrine of professed propriety, disagreeing not only with practice but with practiced ideals. Of late there has been a notable convergence of masculine and feminine, chiefly of the feminine toward the masculine, with mixed-company manners still a bit apart. And so the ancient official doctrine dominates the accent of Sundays and times of trouble, modern this-worldliness presides over mixed-company conversation, together they regulate published discussion, the masculine tradition takes over the talk after a few drinks, and conduct goes its confused way.

The candle has continued to be put out; a social prudery taking the place of theological abhorrence. Indeed, it sometimes goes further: it is in the last hundred-and-fifty years we have the curious assurance that nakedness is primarily an offense against the respectable observer. If the development were to become complete, to be sure—complete along some of its twisted threads —the state of the candle might be indifferent, for the doing itself would tend to indifference.

So far as the put-out candle has stood in the way of this modern change in the ancient code of Christendom, it doubtless has been in part by safeguarding what was unfortunate in that code against the utilitarian cleaning up, but also it has been by safeguarding what is true in the old code against the utilitarian triviality—it has been by symbolizing the mysteriousness and significance of bodily communion. Platonism and Christianity have found their hold on the world and maintained their adequacy for the world largely because of their depth, which has provided for many strata of human experience. They have gone wrong in leaving no real content for their spiritual other-worldliness and in regarding sense as only the food of destruction. Yet, if sense served the devil the devil was real; and to fight sense and the devil was occupation enough for many a godly battler whose sense was strong; content to leave it to God to fill up the vacuum when sense had been at length cast out. Through mysticism (about which the church was never comfortable but which it was wise enough never to rule out), through ritual, through relapses into sensuality (to which the reality of the devil gave a wonderful satisfactoriness and which could still be repented and forgiven), sense obtained positive outlet.

But today only jazz bands are "hot." The bodily giving of oneself and taking of another is too central and potent an intoxication to be reduced, as today society is reducing it, to a pretty sentimentality, a passing amusement, or a vulgar joke. The Orphic elements in human experience are too real to be expunged from existence by an up-to-date rotarian metaphysic or cured by an up-to-date efficiency psychology; even when this is sought in the interest of individual frankness and social cleanness.

And, without probing so far, sheer exuberance of sexual power, catholicity of sensual capacity, bravery of voluptuous imagination, are still gifts to human experience and possibility, are still power. Should utilitarianism contemn power? But though we worship and serve power in kilowatts and dollars, and though we speak sweetly of spiritual power, all we can think when we so speak is of rising inflections on the radio on a Sunday afternoon or over the finger bowls of a "service" luncheon. And so we regard sensual power as something the individual should (not be afraid of in the thankful fear of God, but) be ashamed of, and which society should relegate, sublimate, and minimize for the sake of business regularity and an easy respectability.

But it is easy for rhetoric to tease itself into unfairness to Main Street and Chatauqua. The most obvious change derivable from not putting out the candle would be the straight reversion to animal directness. Let us get rid of restraints, pruderies, romantic involutions, artificial appetites, cultivated wishes and aversions and indifferences, theoretic prohibitions and justifications altogether, and take bodily desire as the simple occasional physiology which residually it is, accept it, gratify it promiscuously as the moment suggests and offers without any further implication, and pass on. Horrid as this may sound, it is not without much sincere goodness only too easy to see and only too easy to envy after the thought of the pain, excess, repression, deprivation, dissoluteness, humbuggery, vulgarity, barrenness, and dishonesty with which the world has been plagued in following even the noblest efforts at the idealization of sex. The dirt which is earth is not dirty—unless it has been shut under man's pavements. We watch animals in their play almost haphazard with one another; we think what we have gained in richness, personality, and meaning; but also we envy them, for we can scarcely play at all, even with wife, husband, lover—indeed it is rather, if at all, with the chance licentiate we play a little and with the chosen beloved we put out the light and are shamefaced and serious, not with the seriousness of

passion but with the seriousness of an ingrained respectability and a worldly prudence, the gift not only of the
world but of twisted gods. Perhaps our gods would have
been less cruel if they had been got in the light of a
more sportive and sceptical hopefulness instead of between the dark sheets of anxious piety.

Yet reversion to the animal is a counsel of defeat and,
after admitting its negative goodness, I doubt if such a
program (aside from its very doubtful possibility psychologically and socially after our actual history) would
be asked for by any but the most extreme pessimist. To
give up hope of accomplishment and discovery is too
great a price to pay for assurance of peace and honesty.

And there is in such a program a fundamental dishonesty, for it refuses to see what is the most important
if not the most stubborn fact of our being, that we are
conscious. Desire is not merely a physiologic uneasiness
to be got rid of, it is an accompaniment, a spur, and a
fulfillment to emotion; emotion an accompaniment,
spur, and fulfillment to consciousness of one's self, one's
otherness, one's communion. It may be said that friendship, social feeling, love may all maintain themselves
without muddying with sex; but, though it would be
unsafe to say there is essentially necessary connection
between love and the mutual giving of bodily pleasure,
it is safe to say that by its essential character this is
supremely and uniquely eligible for the role of expressing that. To rule them apart is to make an insane
sacrifice or an insane denial. And if the two remain

relevant and love is to be anything more than a spasmodic gleam on the crests of gratification, then gratification must undertake a conscious responsibility more than simple physiology. The lighted candle must do more than show indifference to taboo; it must discover perspectives of discrimination.

The lighted candle at once suggests esthetic discrimination. Possibly the program may be found here—in the gratification of all the senses as an end sufficient, justifying itself and making its own requirements both of prodigality and of discipline. Gratification is already the end of the masculine tradition, but, by its focused purpose, especially by its exclusion of sight, it has tended to remain or always return to the very dregs of esthetic—what is added is social and prideful not sensuous. Let the candle stay lighted and the official ethic lose its hold, then men and women both may frankly seek the variety of delight, rediscover and admit the art of the Perfumed Garden, select both partner and technique according to esthetic fitness and the pleasure available, and not merely not put out the candle but multiply and vary its light, call in the sun, and use darkness itself not as a cover but as a novelty in illumination.

This is a most evident possibility, and in this direction an actual trend is most evident. Yet so opposite a development can only with utmost difficulty become complete, pure, or stable. In the supremacy of esthetic criteria in personal relations is much positive gain but also loss and offense to obstinate beliefs and feelings. It is not that it would be of necessity the sheer debauchery which many would at once and only see in it, for in the esthetic of the art of love there would be the same possibility of refinement, elaboration, discovery, and perfecting as in all art, and in its personal conduct there would be the same exaction of restraint, bravery, and management as in all conduct. The conventional attitude toward it on the part of many may be taken as actually necessary; this attitude would push toward reaction and would constantly push the esthetic development itself toward extremes, defiance, a feeling either of its own naughtiness or self-righteousness—toward impurity.

Furthermore, the esthetic development would infringe deeply rooted ideas of the goodness of modesty apart from the goodness of asceticism. And although it would be selective as compared with the masculine tradition, its selectivity would certainly be numerically generous and temporally inconstant, thus directly flouting monogamy and the ideals of loyalty, and herein in conflict with our basic social pattern and of almost the same necessity with our basic economic pattern.

And, indeed, even when viewed as esthetically as a modern citizen can view it, the pretty picture of "paganism" laid out in modern setting either seems a wilful luxury available only for a few, or taken simply and pastorally and apart from practical difficulties still is apt to show itself arty, precious and fragile. I say "available only for a few" not merely because of financial requirements but also because the very nature of the thing leads toward a sort of natural aristocracy, since, though almost everyone has some beauty of body, few there be can claim it without some charity of scrutiny. The arty fragility comes, perhaps, in part from the voices of its premature proclaimers but certainly resides in part in the stubborn feeling of most of us that so invasive a personal function requires a more inclusive government than the purely esthetic.

Must we revenge the exclusion of the esthetic by the exclusion of everything else? If not, can we merely welcome in the esthetic and be content? No; for the candle which welcomes home the esthetic prodigal hopelessly affronts the old moral government of the house, and if the prodigal is to remain at home the father will have to give up his aversion to fatted calves, and if he, too, is not to become a prodigal or leave his home to the prodigal some rescue of moral authority must be made. There is then this possibility; the ideals of modesty and chastity may survive the wreck of asceticism, find a positive basis and reward congenial with esthetic sensuality, and be encouraged not shamed by the now persistent candle. It would be foolish to say this will be; but it might be and the candle may help it to be.

If modesty be the regard for what is intimate rather than the suppression of what is shameful, if it be the reservation of what is prized for its best use and highest enjoyment, if chastity be not mere continence but freedom from dissolute connections and wishes, if modesty and chastity connote the preservation of the flesh to the service of knowledge and love; then modesty and chastity need no help from asceticism and give it none.

Such interpretations of old ideals are not new even in the mollified usage and theory of Christendom when it is not on the high horse of official asceticism; at least the negative side is familiar. The positive, which not only permits but also calls for sensual warmth and enjoyment where modesty rightfully divulges itself, here is the balk. There is the respectable feeling that "passion" is illicit, an animal fact to be made the least of and "sublimated," that true love should approach the "platonic" (*pace* Plato); once let in passion and there is no distinguishing. How may one tell between lustful temptation and loving desire? The test is not so difficult. Is one most of all interested in the other one's consciousness rather than his own, and is he so not through what might be a psychological habit and luxurious refinement but through a real interest in the other person? Yes or no; love or lust.

This is, of course, a matter of infinite degrees; and people differ widely in the easiness and frequency of its occurrence. Moralists and sociologists may well debate. One man's constant meat is another's poison at a dose. But back of social conflictions and personal variations the moral basis is unaffected, and it is a basis which can provide for that practical compromise the institution of marriage, for the discipline of personal reticence and restraint and loyalty, and for the richness of esthetic enjoyment to which indeed it adds the taste of personal communion which once savored leaves sensual pleasure lacking it a dish unsauced—or a sauce without meat.

This is not merely a hope to salvage what may be hung on to in the wreck of old valuations. It is a prospect of better theory and practice which current changes may bring about. The official doctrine has allowed personal attachment but only as a sort of concession to the flesh and the world, not in itself harmful and of further value if it prefigures and leads on to heavenly love. The masculine tradition, on the other hand, likewise allows and talks of personal preferences; but these preferences are still selfish, limited, and largely insincere; for the prevailing social code makes exercise of preference narrow, the masculine lack of interest beyond simple performance makes it unimportant, and the habitual putting out of the candle leaves availability the real test and sufficient test. The more that masculine practice has been confined to its own backstairs game by church asceticism and its social-utilitarian caretaker, the less part has personal attachment played.

The pressure of fact and insight have been upon both church and laity, but only to the accomplishment of "venial sins," expected wild oats, holidays from business or for business. The other-worldly frown which came from Plato's dialectical sophism and Paul's tormenting taboo has continued strong to debase all extravagation from the ethic of the cloister. What a historic irony that the disciple of Socrates, Plato, himself a seer of beauty, and the disciple of Jesus, Paul, himself a destroyer of law, should by their wisdom and their depth have established upon the world their wrongness against the rightness of the less wise and less deep Epicurus!

The theoretic doctrine of the rightfulness of sense and the coextension of gratification and love is itself no new one and has been preached especially in the last couple of generations. Unfortunately, it has figured chiefly as a doctrine of romanticism and has been in conflict with social and economic standards through its accent upon the unlawfulness of marriage and marital relations without love and upon the lawfulness of extra-marital relations with love. But these accented conclusions are unnecessary, for the doctrine properly is a fundamental ethical one of selection in personal relations, a selection upon which a further selection may be imposed with the imposition of further social schemes. Marriage and all the complication of rights involved and the resulting compromises required—all this is a question for further social casuistry.

The socially rebellious romanticism of the doctrine as preached has been chiefly because it has been involved with the romantic thesis of the fleetingness and involuntary factuality of love and in general with that widespread tendency of postromantic thought which makes all human behavior a matter of brutal and unquestionable pushes from the irrational and refuses to give any but an illusory authority to the ancient ideal of foresight and reason. If love be the simple and only occasion for sensual intercourse, and if love be nothing but an irrational sickness which falls upon people and falls off again without their having anything to do about it, then marriage, whether as a deliberate undertaking or as the result of another irrational shove (but still a continuing institution), must be incommensurable with personal ethics, and the only issue is in a socially recognized free love. Our foremost dramatist, in a letter to his wife reminding her of a prenuptial agreement and announcing his intended divorce and remarriage, takes this theory as mutually granted if not self-evident to all intelligent people (intelligence as usual being called in to witness the truth of a theory which denies intelligence all other rights).

Now it is true that love can be so accepted, and it may be true that it would be well if society generally adapted its schedules to fit such acceptance; but the psychology which sees no further possibility is simply false. Love is a brutal possession; it is also in part brought on and determined by one's self, including his thoughts and intentions; it is in part subject to criticism and valuation, and in its continuance it is very largely dependent upon reason—upon what we do to it deliberately or not.

The marriage vow to love is not an absurdity as many now freely assert. We cannot assure continuance but we can do our best, and the vow is the more relevant in that if anything is certain in the psychology of love it is that love will die which is not cared for. We may, of course, either for or against do the opposite of what we try to do, by mistaking the means; or we may accomplish either with great labor yet without being aware of making the effort. But the normal permanence of marriage, to which our economic system is adapted, is not a psychological impossibility or ethical anomaly —however the actuality may turn out.

Reason is not the creator of desire, still less its proper enemy; but reason may now rejoice in desire and now disapprove, and reason may recognize that not all that is permitted and is good in the simplicities of Arcadia is socially expedient. (Our reason has been contrarily taught to disapprove of desire as desire with the concession that not all the fleshlessness of abstract perfection can expediently be required in our material world.) Reason cannot manufacture love, dismiss it offhand, or make sure its stay; but reason and only reason can do all that can be done, and that is not little, for the constancy and integrity of love.

Any complex society will be a thing of many compromises; but there is no essential conflict between the traditional social-economic ideal of stability in personal relations, the traditional moral ideals of loyalty and faithfulness, and the primary ethical values of lusty personal desires and steadfast personal modesty.

I say primary ethical values because a metaphysics of ethics is not lacking if one wish for it. What we find when the ethical question first arises is that we are separate conscious persons existing in a world together with other separate conscious persons. Having made the first choice in favor of consciousness, in favor of go-on rather than go-back, then we may recognize the two primary "virtues," that is the characteristic functions of what we find and choose, as desire, the forward and communicative energy, and modesty, the energy and reticence which maintains the personality which we have accepted. The one is the goodness of that bad lust, the other the goodness of that bad pride, which our negatively anxious fathers in the Middle Ages taught as the roots of all evil. Following close to these first two commandments to enjoy and to refrain, with the recognition of the equal reality of our fellows would come the third and fourth commandments, do good to others and mind your own business.

All other virtues are secondary and derivative, like mercy and industry; or formal and instrumental, like courage and honesty. These last may be said to be first of all, but we are not here concerned with them and need not disturb ourselves whether form is first or last.

Now, the put-out candle has never quenched lust, but it has belittled and falsified the communicative generosity and enfeebled the selective prerogative of desire; it has done nothing to expose the humorlessness of selfish pride, while it has done nothing to help the pride of chastity which saves itself for a richer possession and a richer expenditure. Then there may be hope that the lighted candle may give light at once to the death of asceticism and to the discovery that modesty and chastity are more good, more beautiful, grave, free, and full of gifts than they were when regarded as flat denials. And the bottom of this is that consciousness, the enhancement of consciousness, is the ingredient of all goods, the touchstone of every better; and light, which is the universal symbol of knowledge, is in the doing of the body the condition of knowledge. When we learn that true privacy craves light and not darkness, we may learn that the intimate is not the indecent, and that innocence is not ignorance. I am even willing to repeat the great saying of Socrates, so often ridiculed, virtue is knowledge.

When I am told I should believe in the Virgin Birth —as a Christian and a man of faith—because it is a mystery, I like to reply that I can think of a greater mystery: birth by a father without a mother. I add that I have some textual warrant, for Matthew gives the genealogy of Jesus through Joseph and in the Gospel according to the Hebrews Jesus speaks of "my mother the Holy Spirit." And if I am given time I further add that I do not object to the dogma as a mystery—being a man of faith and finding no impossibility in the belief —but do object to it morally.

The classical world, often called pagan, is supposed to have been unafflicted by our anti-sensualism; and it had many a "virgin birth." Why seems it so particular in us? With the convenient humanness of the Greek gods (the Olympian variety) there was no great mystery in a god fathering a child by mortal woman. And by the same token there is no implied asceticism— quite the contrary: in these divine mixtures the poets have always found the theme of most impassioned ecstasy. It is indeed not fair to speak of these as virgin births. Virginity is scarcely typified in Semele or "wanton Arethusa's azured arms."

In the Christian dogma God is to be sure the father; but God has become "spiritual," God not Jove or Hermes. Thus the emphasis of the doctrine (first doubtless sprung like the others from the wish to glorify the hero's parentage) comes to be not so much on the godhead of the father as on the miracle of the conception and, instead of revering the joy of his making as a divinely magnified joy wherein the god himself gives warrant to the spiritual perspective of sense, we come to take his creation as the avoidance of reality and the denial of pleasure; so that if we are to find good it must be not by the pursuit, exploration, and perfecting of the goods we know but by escape from them.

The Virgin Birth is doubtless to blame that American cooking is poor, that American meals are unenjoyed except as accompaniment for a newspaper or a "deal," and that American digestions are bad. The Virgin Birth makes us dyspeptics and dyspepsia confirms our faith. To be sure indulgence can make us dyspeptics and ruin our faith in pleasure. Is it better to have a true faith falsely falsified or a false faith falsely confirmed?

"Wisdom is got only with experience," we are told.

It should be added, "and then only when experience is got with wisdom." (It is still true that he who "wisely" goes about to have experience loses his pain.) Further, experience is queer: we cannot have without missing. And he who misses may have more than he who has; may have the better experience for the getting of wisdom even in that which he misses, and may have the more experienced experience. If we have the possession of women we miss the lack of them, do we not? It is probable that Simon Stylites had a more unique and intense experience than Don Juan; and that the virgin can come closer to imagining the experience of the successful lover than the other way around.

All this has to do with natural desire, fulfilled and unfulfilled; not with the discovery of objective data. Byrd has that of arctic scenery, Einstein that of relativity theory, the lack of which was no pressing experience to King Alfred. But who speaks of experience and wisdom without reference to desire? He who achieves and enjoys the obvious fruits misses their poignant lack; and if he has certainly a broader more varied more positive and saner basis of acquaintance, he is much the less urged to the use of imagination and analysis; success dulls, the obvious shuts off pursuit of the possible, and if "still achievement lacks a gracious somewhat," he lets it go at "disillusionment." The wisdom of the failure must escape bland unawareness, bitterness, "sublimation," apologetic asceticism and resignationism.

It is easy, for apparent success and apparent failure, to blame things instead of ourselves, to blame things instead of knowing them. All experience is positive, all difference in experience is qualitative. He who stays all night has not merely more of the same but a different experience from him who plays and runs. You have done both? Then you have done differently from him who has done but the one. And he who has never done either knows what neither of you has known.

Of course, I do not mean it is all a matter of indifference; only, in experience you cannot have all and you cannot have none. If you seek a basis of difference it must come in a judgment upon experience from without, not simply in a difference as experience among different experiences which are all homogeneous and all unique. And though success bears in its name its preferableness, in the service of wisdom it has no such prerogative. Nor am I moralizing the "sweet uses of adversity," but talking of simple knowledge.

Truly in one respect the experience of success (or of both success and failure) is best qualified to give advice, as to the technique of success; and it is this which is the most sought wisdom. Right enough it should be so. But it is wearisome that all things in the dialectic of love should be interpreted only as technique. At least, the aim of love is many or questionable: What success? "He is a fool to act so; he might know that is no way to win her." Is it not conceivable he should yet choose so? Must all things done toward the beloved be matters of engineering? At any rate the true lover is too much in love to make it so—unless he become "wise" which may be least wisdom. Even in that smart sense of "wise" it may be "the tigers of wrath are wiser than the horses of instruction."

"Women do not like so and so." Who is this Woman so assuredly known, and in what court is her liking? Worldly Wise is right if I also worldly wise play for a dim success in my one out of "the great majority of cases." Suppose I stake for more? But is it more? Here we need something other than technique: the office of wisdom, coming only by experience. What experience? Experience honestly got in the pursuit of wise ends other than experience, lighted by imagination, dissected by analysis, looked round about by reason. And whence all this wisdom? Whence but from experience?

A paradox and a dilemma? If a paradox no contradiction; and if a dilemma no impassable one; for some do better than others and we know it. Apparently at least one of our hands has hold of our own bootstrap; we need some sort of free grace—or call it luck. And for this there is faith.

"Faith is the belief in what we know is not so." With a little change the old joke becomes true compliment. Faith is the belief in what we know may not be so. And what is there of any general importance of which we do not know this? So far from being opposed to reason, faith, together with skepticism, is its chief need. Not now one, now the other, but both at once. Either without the other is not itself, is ignorant, presumptuous, lazy, and less than moral. The rationalist is one who is willing to take a chance on reason since no one has a choice except between better and worse reason; he is aware that the conclusion of either the fanatic or the scoffer may be true but that their assurance is certainly false.

Do I find the theory of relativity convincing? I will have some need of faith against its difficulties and against common sense. Do I find the classical mechanics still preferable? I will have some need of faith against its hiatus and against new fashion. The atheist will need faith to keep him from God's comfort in privation and terror. He who believes in God must cling to his faith through lazy calms and disillusioning failures. Put it in one case: Do I now love? And heart and head tell me my love is good? Yet it will take all my faith to continue to love against the dead weight of—not any new and better reason but—satiation or despair or time "which wars on all beautiful things."

The great need for faith and difficulty for faith is not, as has been supposed, in the mysteriousness of that which nevertheless we ought to believe, in defiance of reason. How are we to know apart from some reason for it what it is we ought to believe? The need and difficulty for faith lie in the push and pull of motives properly irrelevant as reasons but destructive of that good thing which some reason has bid faith hold to.

On the other hand, more or less of internal mystery

is no bar to faith and is a natural call to it. And so I repeat the mystery of the Virgin Birth does not offend me. It is indeed a feeble enough mystery (especially if the orthodox but uncanonical *virginitas in partu* is left aside). Certainly it is not usual for a human body to be born of woman unknown by man. Yet, I should be more surprised if no such birth had ever occurred than if some had.

Reason is dependent externally and for its consequences upon faith; it is internally not possible without intuition, direct apprehension of simple fact and relational truth and value. But also it needs skepticism. We intuit what we intuit and otherwise we know nothing; the trouble is we have no pure intuition and no assured intuition as to just what part of our perception and acceptance and assertion is the gift of intuition. It may be said that intuition is never mistaken but that we may always be mistaken even though we are not mistaken in saying that some of our knowing is intuitive. So we need faith and have a right to faith in our faith; but faith is a gamble and can bet wrong. Faith is doubtless from the father of all light but our faiths—what we may waste our fine faith upon—are sometimes just proprieties. Or they may be the rebellion of vanity against propriety. Intuition and faith need to be checked on by more commonplace and suspicious underlings— like the "little dog at home" who will know the old woman who had her skirt cut off. But the little dog did not know her, he barked at her: "Lawk a mussy on us, this is none of I." The helpful underlings, helpful because commonplace and suspicious, may be mistaken in differentiating their betters and impostors. But they often—normally, like the little dog—save us. There is no more proper cry than that of the father of the possessed boy: "Lord, I believe; help thou mine unbelief." But needful too is its other: "Lord, I doubt; help thou my belief." How many of our faiths are idols of the tribe, the cave, the market place, the theater, tenacity, propriety? "Their spirit," Zarathustra said, "is imprisoned in their good conscience."

In the Christian incarnation, I think, God took on the human liability to doubt, ability to doubt, need to doubt. "Know ye not I must be about my Father's business?" Did Jesus know it—the reality of the call, what the call was, his ability to give the answer in the world about him and ahead of him? He had his grounds and on them his faith; but he also knew he might be variously wrong or might variously fail. And this is so in degree with all those who try for anything beyond the ordinary. One assumes the extra risk that he be presumptuously wrong, mistaken in his credentials this way and that. But to avoid that risk is to risk being supremely wrong. Of those who try, and thereby fail of lesser things they could do, many are wrong; but I should say most of these are also right in having some decent ground for taking the braver risk. The social discountenancing or the friendly hesitancy to countenance is like the parental love which prefers the safer road for the child.

I think luck as well as freedom must be counted in the salvation of man as well as in the fall; and not merely as of individuals but as of the cosmic drama. I believe that luck should be counted in the story of Jesus. God may have known He had a good bet but He had to wait for the finish. He may even have made the bet before. It not only must have been in the power of Jesus to fail, it may have been in the power of events to prevent his achievement. I say "must" of the necessary freedom for the goodness and the sacrifice; "may" of the continuing normal noninterference of God in the world in which Jesus achieved his goodness and his sacrifice. It seems to me essential to the true divinity of this man that he should be truly man not only in his body but also in his knowledge and his doubts and fears.

Index